Synthesis Lectures on Renewable Energy Technologies

The series, Synthesis Lectures on Renewable Energy Technologies publishes concise books, focused on technologies that harness energy from naturally occurring sources, such as sunlight, wind, water, geothermal heat, and biofuels from organic materials. These renewable energy technologies play a crucial role in transitioning away from fossil fuels, helping to mitigate the effects of climate change, and promoting a sustainable energy supply.

Nawel Mensia · Mourad Talbi

Innovative Solutions in Simulation, Modelling and Control of Photovoltaic Systems

 Springer

Nawel Mensia
Photovoltaic Laboratory (LPV)
CRTEn Borj Cedria
Hammam Lif, Tunisia

Mourad Talbi
Laboratoire de maitrise de l'Energie Éolienne
Et de Valorisation Énergétique des Déchets
(LMEEVED)
CRTEn Borj Cedria
Hammam Lif, Tunisia

ISSN 2690-5000 ISSN 2690-5019 (electronic)
Synthesis Lectures on Renewable Energy Technologies
ISBN 978-3-031-94259-4 ISBN 978-3-031-94260-0 (eBook)
https://doi.org/10.1007/978-3-031-94260-0

© The Editor(s) (if applicable) and The Author(s), under exclusive license to Springer
Nature Switzerland AG 2026

This work is subject to copyright. All rights are solely and exclusively licensed by the Publisher, whether the whole or part of the material is concerned, specifically the rights of translation, reprinting, reuse of illustrations, recitation, broadcasting, reproduction on microfilms or in any other physical way, and transmission or information storage and retrieval, electronic adaptation, computer software, or by similar or dissimilar methodology now known or hereafter developed.
The use of general descriptive names, registered names, trademarks, service marks, etc. in this publication does not imply, even in the absence of a specific statement, that such names are exempt from the relevant protective laws and regulations and therefore free for general use.
The publisher, the authors and the editors are safe to assume that the advice and information in this book are believed to be true and accurate at the date of publication. Neither the publisher nor the authors or the editors give a warranty, expressed or implied, with respect to the material contained herein or for any errors or omissions that may have been made. The publisher remains neutral with regard to jurisdictional claims in published maps and institutional affiliations.

This Springer imprint is published by the registered company Springer Nature Switzerland AG
The registered company address is: Gewerbestrasse 11, 6330 Cham, Switzerland

If disposing of this product, please recycle the paper.

Perface

The increasing demand for clean and sustainable energy is driving the need for innovative approaches in modeling and controlling photovoltaic (PV) systems. This book, titled Innovative Solutions in Modelling and Control of Photovoltaic (PV) Systems, presents a solid and flexible framework aimed at enhancing PV efficiency in real-world applications through high-performance control strategies. The book is organized into three detailed chapters. It begins with a comparative analysis of traditional Maximum Power Point Tracking (MPPT) algorithms—Perturb & Observe and Incremental Conductance. These techniques are meticulously examined using MATLAB/SIMULINK simulations under both standard and varying climatic conditions.

In the second chapter, a high-performance and adaptable modeling method is introduced, based on a multi-model approach specifically designed for PV water-pumping applications. This methodology represents the PV system's behavior under varying environmental conditions through a multi-model structure established using a convex polytopic transformation. The structure comprises eight linear local models, assigned using nonlinear weighting factors to ensure a dynamic and accurate representation of system performance. The simulation results highlight the model's accuracy in capturing system behavior and optimizing energy output across a range of operating conditions.

The final chapter builds on this methodology by introducing an adaptive proportional-integral controller developed within the same multi-model framework. This controller, designed for real-time implementation, showcases outstanding performance in achieving quick, stable, and precise MPP tracking while canceling output oscillations. Its scalability and compatibility with embedded platforms make it a practical and forward-thinking solution for renewable energy systems. Together, these advancements provide researchers and engineers with powerful tools to enhance the reliability, responsiveness, and efficiency of PV systems.

Hammam Lif, Tunisia

Nawel Mensia
Mourad Talbi

Competing Interests The authors have no competing interests to declare that are relevant to the content of this manuscript.

Contents

1 **Modeling and Simulations of Photovoltaic Systems** 1
   1.1 Introduction 1
   1.2 Perturb and Observe Controller (P&O) 2
   1.3 Incremental Conductance Controller (IC) 3
   1.4 Modelling Under Matlab/Simulink of Photovoltaic Systems 5
   1.5 Results and Discussions 8
   1.6 Conclusion 25
   References 25

2 **Photovoltaic System's Modelling Based on Polytopic Transformation** 27
   2.1 Introduction 27
   2.2 Photovoltaic Water Pumping System 28
   2.3 State Model Elaboration 31
   2.4 Polytopic Transformation of the Water Pumping Photovoltaic System's State Model 33
      2.4.1 Multi-model Modeling 33
      2.4.2 Model's Base Establishment 34
   2.5 Results and Discussions 38
   2.6 Conclusion 42
   References 43

3 **Effective PI Controller for Maximum Power Point Tracking of a Photovoltaic System** 45
   3.1 Introduction 45
   3.2 Representation of the PV System State Model 47
   3.3 PV System State Model Transformation Using Polytopic Transformation 49

3.4	PI Controller Development	51
	3.4.1 dP/dV Feedback Control Strategy	51
	3.4.2 Partial PI Controller Synthesis	52
	3.4.3 Establishment of the PI Controller for the PV System	55
	3.4.4 Stability of the PI Controller	56
3.5	Results and Discussions	60
	3.5.1 Model Simulation	62
	3.5.2 Controller Simulation	65
3.6	Conclusion	68
References		69

List of Figures

Fig. 1.1	The flow chart of a P&O controller	3
Fig. 1.2	MPP deviation with P&O controller under rapid changes of insolation	4
Fig. 1.3	MPPT Controller employing incremental conductance (IC) method	6
Fig. 1.4	MPPT controller employing the modification IC technique [11]	7
Fig. 1.5	The first PV system using IC controller with STC conditions ($T = 25\ °C.\ G = 1000\ W/m^2$)	8
Fig. 1.6	The second PV system using P&O controller with STC conditions ($T = 25\ °C.\ G = 1000\ W/m^2$)	9
Fig. 1.7	The block diagram of IC controller [14]	9
Fig. 1.8	The subsystem of computing the load current, I [15]	10
Fig. 1.9	The subsystem of computing the thermal Voltage, V_t [15]	10
Fig. 1.10	The subsystem of computing the reversed saturation current, I_s [15]	11
Fig. 1.11	The subsystem of computing the diode current, I_d [15]	11
Fig. 1.12	The subsystem of computing the reversed saturation current at Top, I_{rs} [15]	12
Fig. 1.13	The subsystem of computing the phase current, I_{ph} [15]	12
Fig. 1.14	The subsystem of computing the shunt current, I_{sh} [15]	12
Fig. 1.15	The block diagram of P&O controller [16]	13
Fig. 1.16	The curve of temporal variation of the power provided by the used GPV (green box in Figs. 1.5 and 1.6), obtained in case of STC (the temperature, $T = 25\ °C$ and the insolation, $G = 1000\ Watt/m^2$)	13
Fig. 1.17	A zoomed part of the curve of temporal variation of the power provided by the used PV panel (Fig. 1.16)	14
Fig. 1.18	The P–V characteristic obtained in case of STC (the temperature, $T = 25\ °C$ and the Insolation, $G = 1000\ W/m^2$)	14
Fig. 1.19	The curve of temporal variation of the insolation (Fig. 1.6)	15

Fig. 1.20	The obtained curve of temporal variation of the power provided by the used GPV (green box in Fig. 1.6)	15
Fig. 1.21	A zoomed part of zone (**A**) of the obtained curve of temporal variation of the Power provided by the used GPV (green box in Figs. 1.5 and 1.6)	16
Fig. 1.22	A zoomed part of zone (**B**) of the obtained curve of temporal variation of the power provided by the used GPV (green box in Figs. 1.5 and 1.6)	16
Fig. 1.23	The P–V characteristic obtained in case where the temperature, $T = 25\ °C$ and the Insolation, $G = 800\ W/m^2$	17
Fig. 1.24	A zoomed part of zone (**C**) of the obtained curve of temporal variation of the power provided by the used GPV (green box in Figs. 1.5 and 1.6)	18
Fig. 1.25	The P–V characteristic obtained in case where the temperature, $T = 25\ °C$ and the Insolation, $G = 700\ W/m^2$	19
Fig. 1.26	A zoomed part of zone (**D**) of the obtained curve of temporal variation of the Power provided by the used GPV (green box in Figs. 1.5 and 1.6)	20
Fig. 1.27	The first PV system using IC controller with climatic conditions variables over time	21
Fig. 1.28	**a** the curve of temporal variation of the insolation (Fig. 1.27), **b** the curve of temporal variation of the temperature (Fig. 1.27)	22
Fig. 1.29	The curve of temporal variation of the power provided by the used GPV (green box in Figs. 1.5 and 1.6), obtained in case where the climatic conditions are variables over time (Fig. 1.28)	22
Fig. 1.30	A zoomed part of zone (**C**) of the obtained curve of temporal variation of the power provided by the used GPV (green box in Figs. 1.5 and 1.6)	23
Fig. 1.31	P–V characteristic obtained in case where the insolation, $G = 500\ W/m^2$ and the temperature, $T = 0\ °C$	23
Fig. 1.32	A zoomed part of zone (**B**) of the obtained curve of temporal variation of the power provided by the used GPV (green box in Figs. 1.5 and 1.6)	24
Fig. 1.33	P–V characteristic obtained in case where the insolation, $G = 500\ W/m^2$ and the temperature, $T = 50\ °C$	24
Fig. 2.1	Water pumping PV system	29
Fig. 2.2	Equivalent circuit diagram of the photovoltaic system	29
Fig. 2.3	Electrical representation of the single diode model	31
Fig. 2.4	Evolution of the Photovoltaic generator voltage V_p and the panel voltage V_{pm} for temperature varying from 0 to 70 °C and illuminances varying from 200 W/m² to 1000 W/m²	40

Fig. 2.5	Progression of the error between the panel voltage and the model's voltage V_{pm} across a range of temperatures from 0 °C to 70 °C and illuminances from 200 W/m² to 1000 W/m²	40
Fig. 2.6	Evolution of the error between the motor current I_m and the model's current I_{mm} across a range of temperatures from 0 °C to 70 °C and illuminances from 200 W/m² to 1000 W/m²	41
Fig. 2.7	Evolution of the error between the motor rotational speed ω and the model's rotational speed ω_m across a range of temperatures from 0 °C to 70 °C and illuminances from 200 W/m² to 1000 W/m²	41
Fig. 3.1	Circuit diagram of the studied system	47
Fig. 3.2	Block design of the dP/dV MPPT control	54
Fig. 3.3	Control loop of Q_i	54
Fig. 3.4	Control loop of I_{Li}	54
Fig. 3.5	Block diagram of the developed command strategy	57
Fig. 3.6	Current-voltage characteristics (Fig. 3.6a) and power-voltage characteristics (Fig. 3.6b) under various irradiation and fixed temperature T = 25 °C	62
Fig. 3.7	Current-voltage characteristics (Fig. 3.7a) and power-voltage characteristics (Fig. 3.7b) under various temperature and fixed irradiation G = 1000 W/m²	62
Fig. 3.8	Variation of the error between the PVG voltage and the model voltage across a range of irradiance from 200 to 1000 W/m² and temperature from 0 to 70 °C	63
Fig. 3.9	Evolution of the error between the current I_L and the model's corresponding current I_{mL} across a range of temperatures from 0 to 70 °C and irradiances from 200 to 1000 W/m²	64
Fig. 3.10	Evolution of the error between the voltage V_{DC} and the model's corresponding voltage V_{mDC} across a range of temperatures from 0 to 70 °C and irradiances from 200 to 1000 W/m²	64
Fig. 3.11	Evolution of the PV system voltage, power, current, and control signal under standard conditions of temperature and irradiance	65
Fig. 3.12	Evolutions of the PV system power, the controller gains, and the control signal under climatic conditions change from (T = 25 °C; G = 1000 W/m²) to (T = 20 °C; G = 800 W/m²)	66
Fig. 3.13	Power variation of the PV system controlled by the MPPT controller based on P&O method (Pink curve) and that of the same system controlled by the proposed PI controller (Blue curve) in the case of climatic conditions change from (T = 25 °C; G = 1000 W/m²) to (T = 20 °C; G = 800 W/m²), and too zoomed parts A and B of these curves	67

List of Tables

Table 2.1	Parameters of Eqs. 2.1 and 2.2	30
Table 2.2	Parameters of the PV cell equations	32
Table 2.3	Characteristics of the PV system	39
Table 3.1	parameters of the PV cell equations	48
Table 3.2	Parameters of the PV module SPM (P) 250W for (STC) and (NOCT) conditions	60
Table 3.3	Parameters of the buck convertor	63

Modeling and Simulations of Photovoltaic Systems

1.1 Introduction

The increased worries of nature and rising demand for energy as well as the world attention to global warming have induced thinking for finding clean resources for energy, renewable energy that varies as wind, geothermal, hydro, solar and bioenergy. Photovoltaic (PV) energy is one of such energy resources, that its importance has increased due to its inexhaustible nature, cleanness, scalability in power and low maintenance required [1, 2].

There are different reasons that impact the amount of power produced by the PV array like atmospheric temperature, solar irradiation, the dust accumulated at the surface of the panel and the configuration of the electrical connection of the panels in the array. Significantly, power extraction and the efficacy of the PV system are influenced by changing weather and Partial Shading Conditions (PSC) [1, 2]. By employing numerous techniques, the effect of the factors above on maximum power generation can be reduced; MPPT methods are one of these approaches, it has been utilized for extracting the maximum power and enhancing the efficiency of the PV system. MPPT methods were implemented in both of PCS and uniform insulations [3].

Diversities of MPPT controllers were employed and tested; the most common are fractional short circuit current, Perturb and Observe (P&O), fractional open circuit voltage [4], Hill Climbing (HC) controller [1], and Incremental Conductance (IC) [4]. Disadvantages of some controllers, such as the difficulty of implementation, overlook of the MPP during partial shading, as well as low tracking speed, have made the development of traditional methods and relying on other techniques that use Artificial Intelligence (AI) is crucial for making the employment of the PV system more efficient.

© The Author(s), under exclusive license to Springer Nature Switzerland AG 2026
N. Mensia and M. Talbi, *Innovative Solutions in Simulation, Modelling and Control of Photovoltaic Systems*, Synthesis Lectures on Renewable Energy Technologies,
https://doi.org/10.1007/978-3-031-94260-0_1

P&O controller is a widely employed traditional method for tracking MPP that is installed in a commercial PV controller, thanks to its simplicity. Though it has two disadvantages, these include the oscillation in the vicinity of MPP, and which results in unending oscillation in output power. Therefore, reduced energy and lower efficacy will yield. The second disadvantage is the divergence from MPP in case where a sudden change in weather conditions occurs and will fail to track MPP, which leads to loss of energy [1]. Moreover, for obtaining a precise MPPT controller, soft computing innovations like Fuzzy Logic (FL) and Adaptive Neuro-Fuzzy Inference System (ANFIS) as an intelligent method are chosen due to their ability to deal with the non-linear and non-exact mathematical model [3].

The remainder of this chapter is organized as follows: in Sect. 1.2, we will deal with P&O. In Sect. 1.3, we will deal with IC. In Sect. 1.4, we will present the modelling of some PV systems and these PV systems modelling is performed under MATLAB/SIMULINK. Finally, we will conclude in Sect. 1.5.

1.2 Perturb and Observe Controller (P&O)

The P&O [1, 5, 6], sometimes named Hill climbing technique as the most well-known MPPT method. P&O is extensively employed as it is the simplest technique among all MPPT ones. P&O is just measuring the PV's terminal voltage and output current, from which the actual power can be computed and varying the duty cycle of the DC-DC converter is performed until the MPP is achieved.

As the name of the P&O technique states, the process is starting by operating the DC-DC converter with the initial set duty cycle and then starts increasing the duty cycle with a certain step width (user defined), and the power is observed with the adding of each step. In case where at a certain point the power gets less than its previous value that means that the duty cycle should get one step in the opposite direction, in other terms getting to the MPP again and so on... (Fig. 1.1).

With this technique, the operating voltage V is perturbed with every MPPT cycle. As soon as the MPP is reached, V will oscillate around the ideal operating voltage. This causes a power loss that depends on the step width of a single perturbation, in other terms, the larger the step, the larger the oscillations around the voltage of maximum power, and vice versa [7]. In case where the step width is large, the MPPT technique will respond quickly to sudden changes in operating conditions with the tradeoff of increased losses under stable or slowly changing conditions. In case where the step width is very small, the losses under stable or slowly changing conditions will be reduced, but the system will be only able to respond very slowly to rapid changes in both temperature and insolation. The value of ideal step width is system dependent and should be experimentally chosen.

One disadvantage of the P&O controller is in case of sudden insolation increasing, the P&O react as in case where the increase happened as a result of the previous perturbation

1.3 Incremental Conductance Controller (IC)

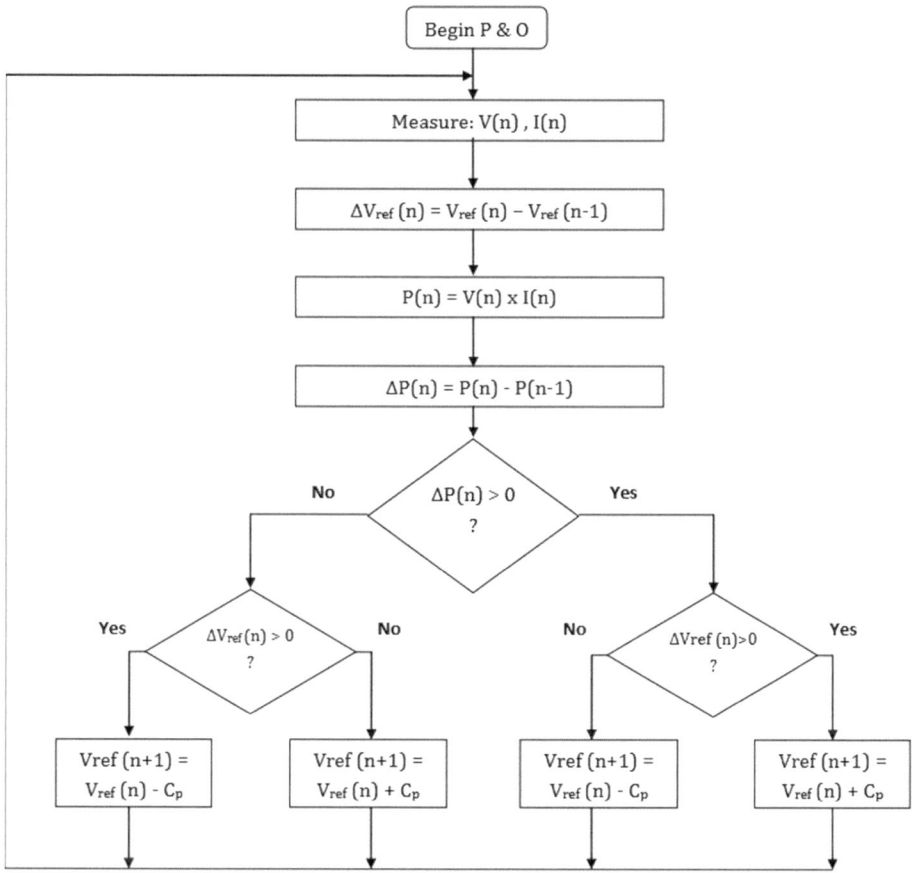

Fig. 1.1 The flow chart of a P&O controller

of the array operating voltage [7, 8]. Consequently, the next operation will be in the same direction as the previous one which can be the opposite direction of maximum power. Figure 1.2 shows that the continuous perturbation in one direction leads to an operating voltage far from the MPP voltage. In case where the insolation change decreases or stops, the MPPT gets back to its normal behavior (Fig. 1.2).

1.3 Incremental Conductance Controller (IC)

The working process of the Incremental Conductance (IC) controller is guided by the incremental conduction ($\Delta I/\Delta V$) of the characteristics of Photovoltaic (PV), which serve for detecting the slope of the P-V characteristic curve ($\Delta P/\Delta V$) [22]. At the MPP point,

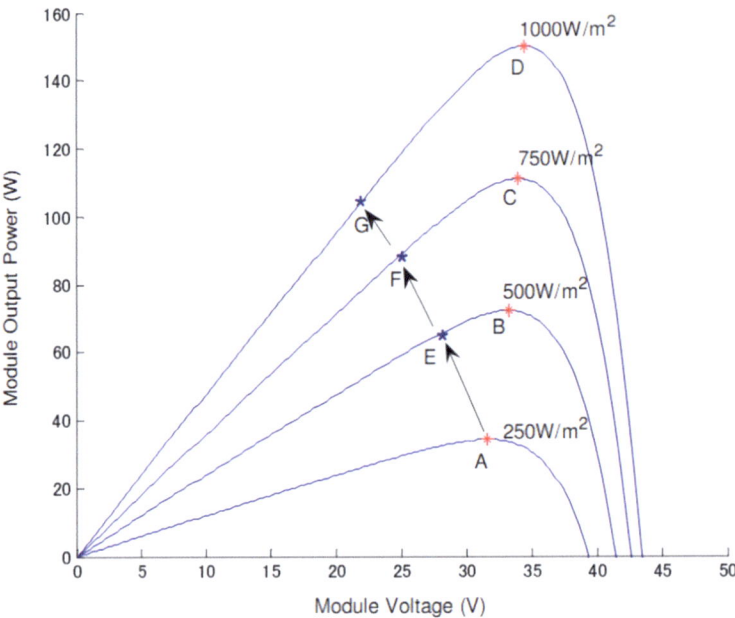

Fig. 1.2 MPP deviation with P&O controller under rapid changes of insolation

this slope value will be zero, large from zero if on the left side of the MPP, and small from zero if on the right side. In case where the slope encountered is negative, then the operating point is shifted to the left by the controller by reducing the PV array voltage value, and vice versa. This happens until the slope becomes zero, and the adjustment of the voltage value will stop, indicating that the MPP is reached [9]. Mathematically, it can be seen in Eqs. (1.1–1.3) [9–11].

$$\frac{\Delta P}{\Delta V} = 0 \text{ at the MPP (Maximum Power Point)} \quad (1.1)$$

$$\frac{\Delta P}{\Delta V} > 0 \text{ at the left position of the MPP} \quad (1.2)$$

$$\frac{\Delta P}{\Delta V} < 0 \text{ at the right position of the MPP} \quad (1.3)$$

The Eq. (1.1) can be developed leading to Eqs. (1.4), (1.5), (1.6) and (1.7) which are stated as follows:

$$\frac{\Delta P}{\Delta V} = \frac{\Delta (V \Delta I)}{\Delta V} = I + V \left(\frac{\Delta I}{\Delta V}\right) = 0 \quad (1.4)$$

Then we have:

$$\frac{\Delta I}{\Delta V} = -\frac{I}{V} \text{ at the MPP} \qquad (1.5)$$

$$\frac{\Delta I}{\Delta V} > -\frac{I}{V} \text{ at the left position of the MPP} \qquad (1.6)$$

$$\frac{\Delta I}{\Delta V} < -\frac{I}{V} \text{ at the right position of the MPP} \qquad (1.7)$$

The Incremental Conductance (IC) controller, as illustrated at Fig. 1.3, is a traditional controller extensively employed nowadays due to its simple characteristics and cheaper financing and application [12, 13].

This IC controller is efficient for tracking MPP under uniform insolation conditions. Though, it is not able to work well in case where there is a rapid changing in insolation and causes oscillations around MPP at a steady state. This controller responds imprecisely to the change of the first step in the task cycle of the converter during increased insolation and the time for finding MPP is also longer. Lying on those problems, it was proposed for modifying the IC controller [11]. This controller aims to quickly track the location of the MPP or optimal operating point on changes in insolation values, particularly for insolation changes that happen suddenly. Figure 1.4 illustrates the modified IC controller [11].

1.4 Modelling Under Matlab/Simulink of Photovoltaic Systems

In this section we will deal with the modelling of two Photovoltaic (PV) systems where in the first one we use a P&O controller and in the second one we apply an IC command. This modelling is performed under Matlab/Simulink. In Fig. 1.4 the first PV system is illustrated and in Fig. 1.5 the second PV system is illustrated.

As illustrated in Fig. 1.5, the first PV system is composed of a PV panel (box in green color), a Boost DC-DC converter and a resistive load. This DC-DC converter is controlled throughout its MOSFET by a Pulse Width Modulation signal named D. . The latter is the output of the subsystem of the IC controller (Fig. 1.5). The inputs of this subsystem are the PV panel current, Ipv and the PV panel voltage, Vpv. This subsystem is illustrated in Fig. 1.7.

The subsystem of the PV panel (box in green color) used in this PV system (Fig. 1.5) is composed by different subsystems illustrated at Figs. 1.8, 1.9, 1.10, 1.11, 1.12, 1.13, 1.14 and 1.15 [15]. These subsystems are listed as follows:

The subsystem of computing the load current, I:

$$I = N_p I_{ph} - I_d - I_{sh} \qquad (1.8)$$

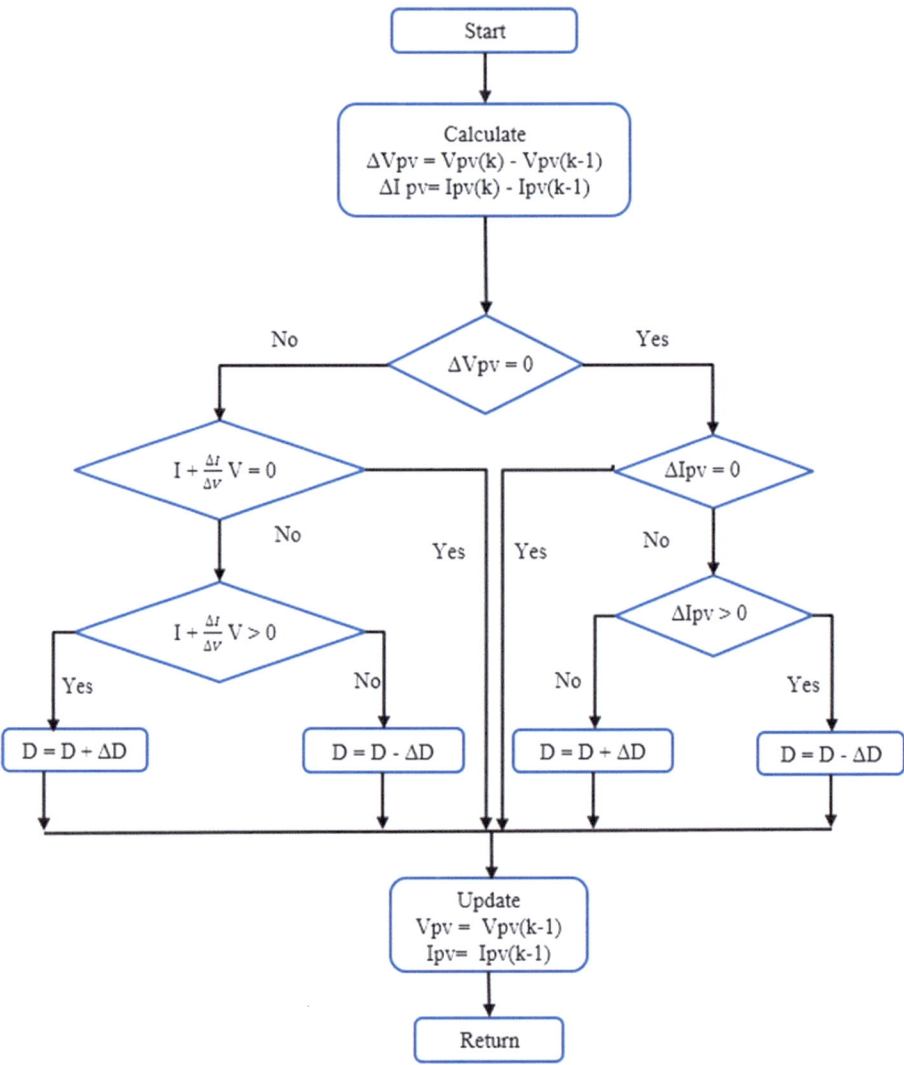

Fig. 1.3 MPPT Controller employing incremental conductance (IC) method

The subsystem of computing the thermal Voltage, V_t,

$$V_t = \frac{k \Delta Top}{q} \tag{1.9}$$

1.4 Modelling Under Matlab/Simulink of Photovoltaic Systems

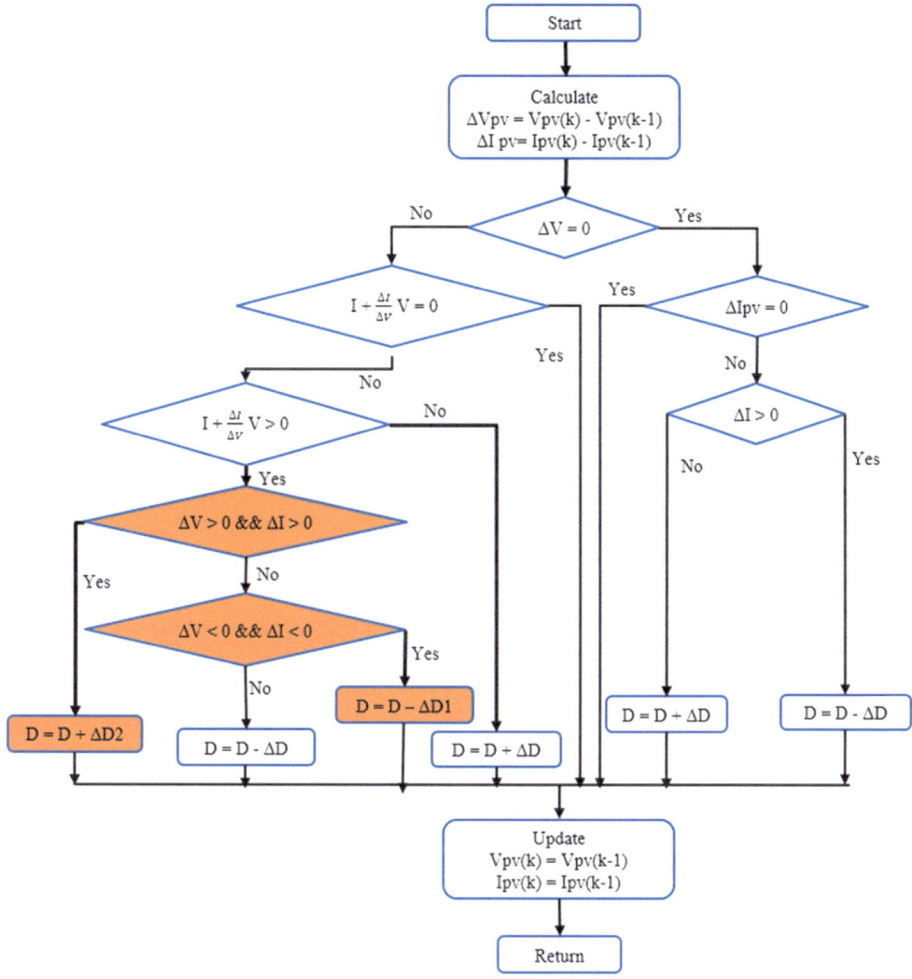

Fig. 1.4 MPPT controller employing the modification IC technique [11]

The subsystem of computing the Reversed saturation Current, I_s,
The subsystem of computing the the Diode Current, I_d,
The subsystem of computing the Reversed saturation Current at Top, I_{rs},
The subsystem of computing the Phase Current, I_{ph},
The subsystem of computing the Shunt Current, I_{sh},

As illustrated in Fig. 1.6, the second PV system is composed of the same PV panel (box in green color) used in the first PV system (Fig. 1.5), a Buck-Boost DC-DC converter (box in pink color in Fig. 1.6) and a resistive load. This DC-DC converter is controlled by the duty cycle which is the output of the subsystem of the P&O controller. The inputs

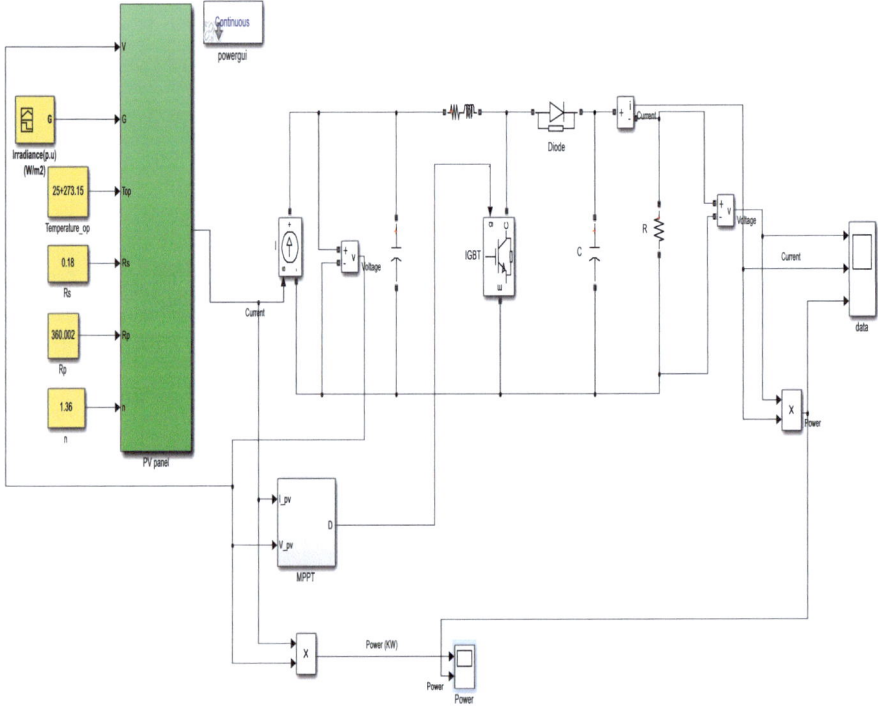

Fig. 1.5 The first PV system using IC controller with STC conditions ($T = 25\ °C$. $G = 1000\ W/m^2$)

of this P&O subsystem are the PV panel current, I_pv and the PV panel voltage, V_pv. This sub-system is illustrated at Fig. 1.15.

1.5 Results and Discussions

In this section, we present the different curves obtained by simulations of the two PV systems presented previously (Figs. 1.5 and 1.6). These curves are the temporal variations of the Power produced by the used GPV (the box in green color in both Figs. 1.5 and 1.6). These curves are obtained in case of STC (the temperature, $T = 25\ °C$ and the insolation, $G = 1000\ Watt/m^2$) and in case where the climatic conditions are variables over time.

Figure 1.16 shows the temporal variation curve of the power produced by the GPV and the temporal variation curve of the power in the output of the DC-DC converter in the second PV system (Fig. 1.6). These curves are obtained in case of STC.

1.5 Results and Discussions

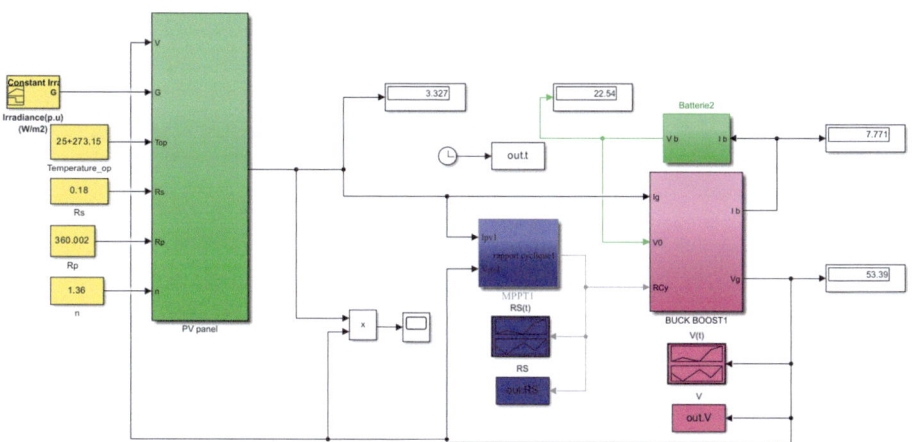

Fig. 1.6 The second PV system using P&O controller with STC conditions ($T = 25$ °C. $G = 1000\ W/m^2$)

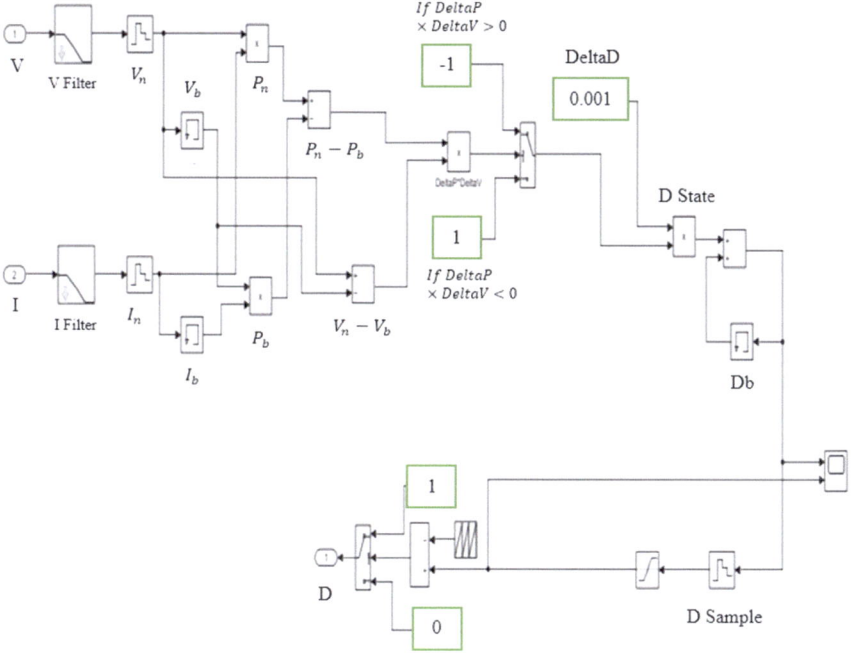

Fig. 1.7 The block diagram of IC controller [14]

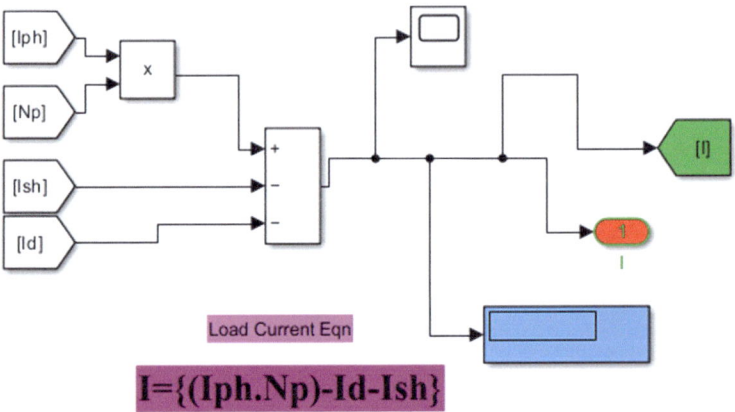

Fig. 1.8 The subsystem of computing the load current, I [15]

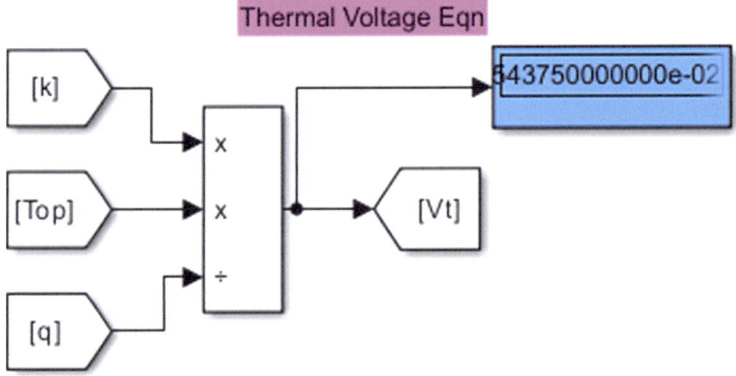

Fig. 1.9 The subsystem of computing the thermal Voltage, V_t [15]

Figures 1.16, 1.17 and 1.18 show that in case of STC (the temperature, $T = 25\ °C$ and the Insolation, $G = 1000W/m^2$), the used P&O controller (Figs. 1.6 and 1.15) permits to track the MPP efficiently.

In this section, we also tested the performance the second PV system (Fig. 1.6) by making insolation variable through time (Fig. 1.19).

According to Figs. 1.20 and 1.21, in case where we take the zone **A** of the curve (Fig. 1.20), the used P&O controller (Figs. 1.6 and 1.15) permits to track the MPP efficiently.

According to Figs. 1.22 and 1.23, in case where we take the zone **B** of the curve (Fig. 1.20), the used P&O controller (Figs. 1.6 and 1.15) permits to track the MPP efficiently.

1.5 Results and Discussions

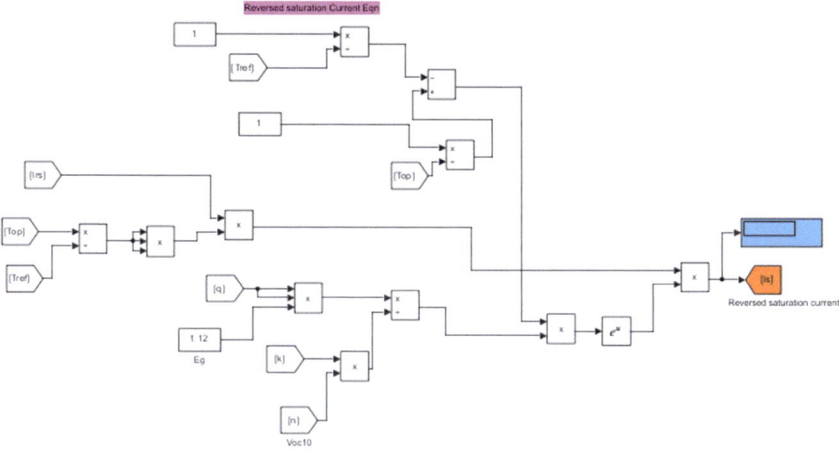

Fig. 1.10 The subsystem of computing the reversed saturation current, I_s [15]

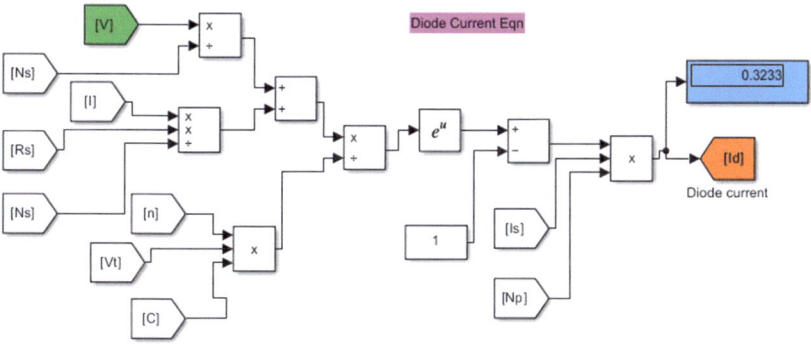

Fig. 1.11 The subsystem of computing the diode current, I_d [15]

According to Figs. 1.24 and 1.25, in case where we take the zone **C** of the curve (Fig. 1.20), the used P&O controller (Figs. 1.6 and 1.15) permits to track the MPP efficiently.

According to Figs. 1.24 and 1.18, in case where we take the zone **D** of the curve (Fig. 1.20), the used P&O controller (Figs. 1.6 and 1.15) permits to track the MPP efficiently (Fig. 1.26).

In summary, the P&O controller permits to track the MPP and this in case where the insolation is variable over time.

According to Figs. 1.29, 1.30, 1.31, 1.32 and 1.33, the IC controller permits to tracking of the MPP, and this is in case where the climatic conditions are variable over time (Fig. 1.28).

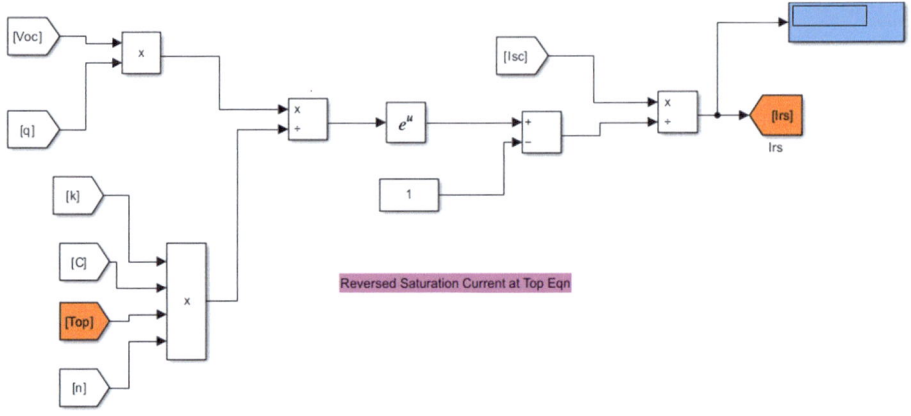

Fig. 1.12 The subsystem of computing the reversed saturation current at Top, I_{rs} [15]

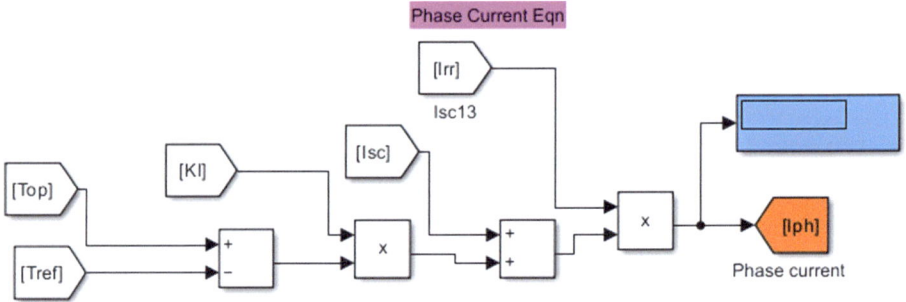

Fig. 1.13 The subsystem of computing the phase current, I_{ph} [15]

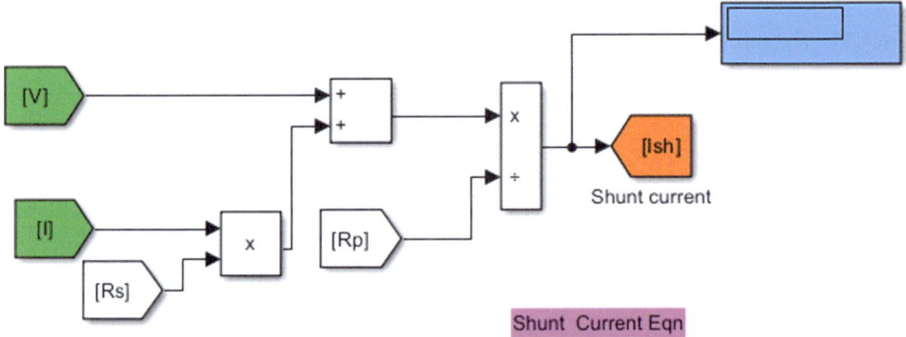

Fig. 1.14 The subsystem of computing the shunt current, I_{sh} [15]

1.5 Results and Discussions

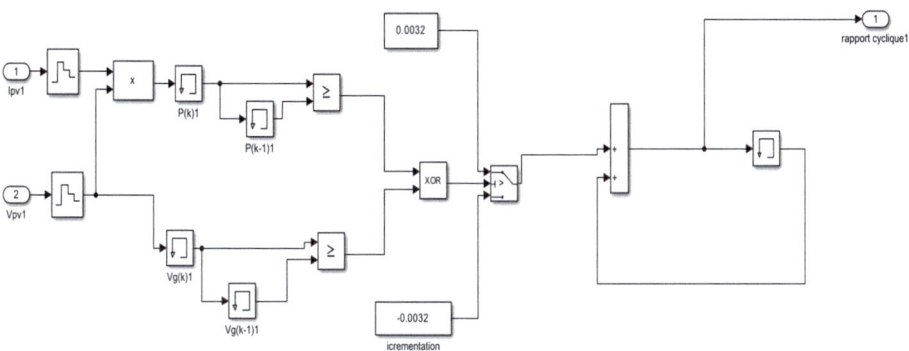

Fig. 1.15 The block diagram of P&O controller [16]

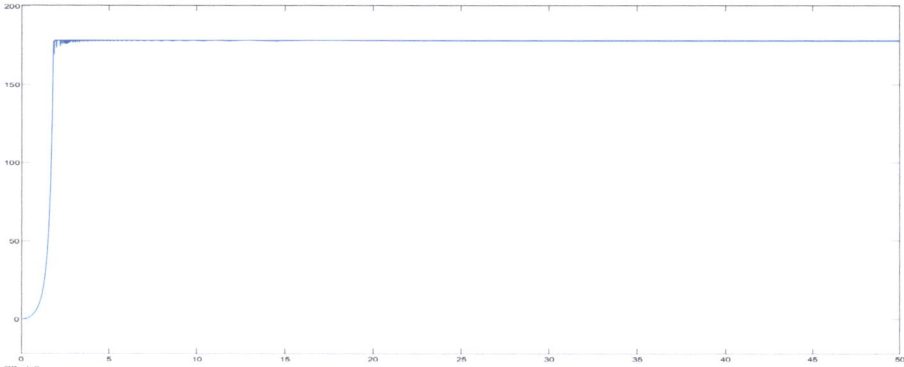

Fig. 1.16 The curve of temporal variation of the power provided by the used GPV (green box in Figs. 1.5 and 1.6), obtained in case of STC (the temperature, $T = 25\ °C$ and the insolation, $G = 1000\ Watt/m^2$)

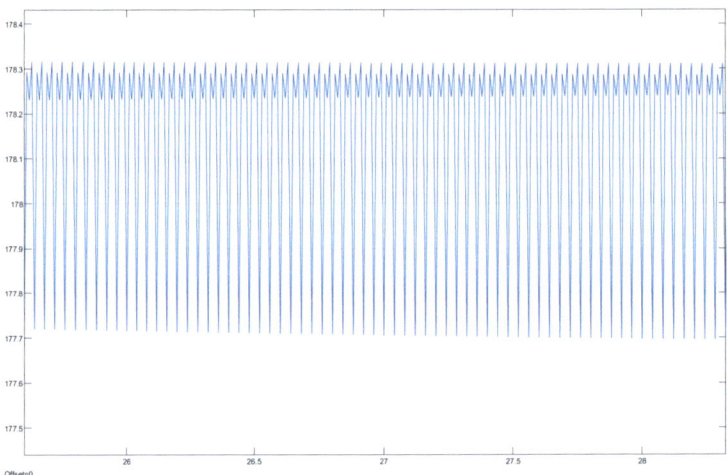

Fig. 1.17 A zoomed part of the curve of temporal variation of the power provided by the used PV panel (Fig. 1.16)

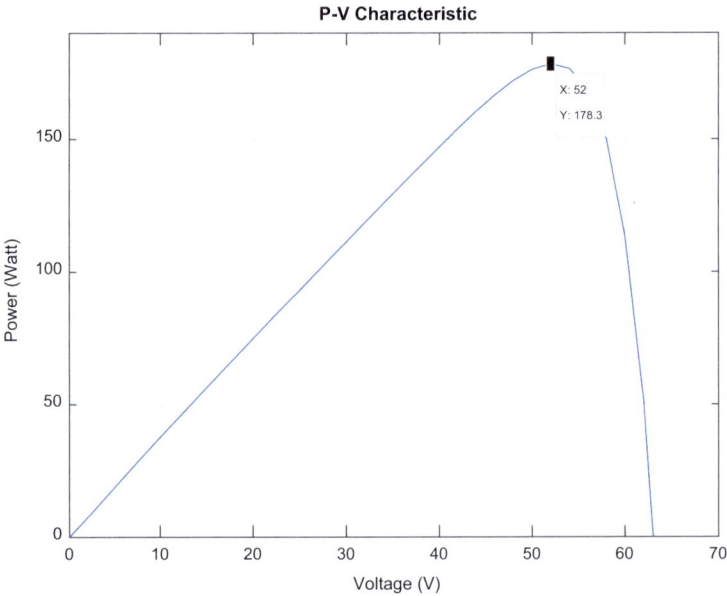

Fig. 1.18 The P–V characteristic obtained in case of STC (the temperature, $T = 25\ °C$ and the Insolation, $G = 1000\ W/m^2$)

1.5 Results and Discussions

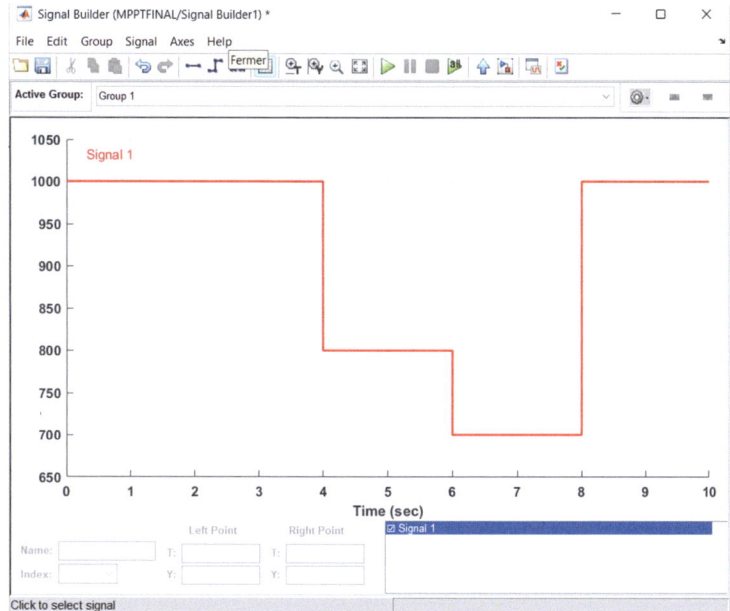

Fig. 1.19 The curve of temporal variation of the insolation (Fig. 1.6)

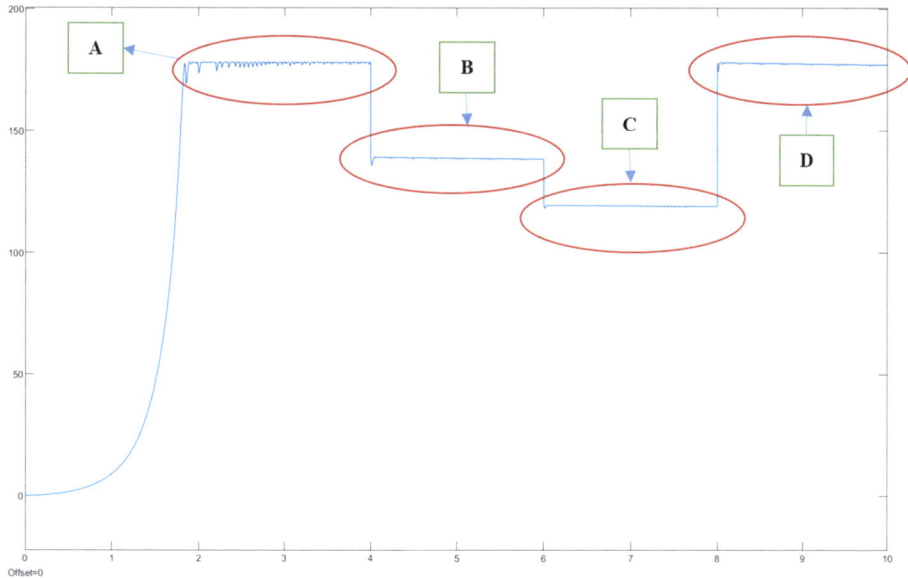

Fig. 1.20 The obtained curve of temporal variation of the power provided by the used GPV (green box in Fig. 1.6)

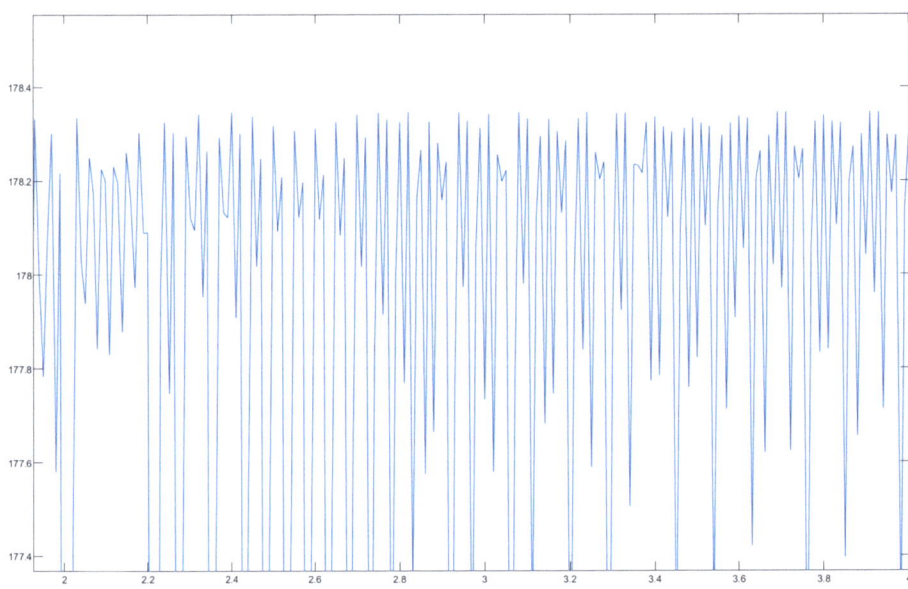

Fig. 1.21 A zoomed part of zone (**A**) of the obtained curve of temporal variation of the Power provided by the used GPV (green box in Figs. 1.5 and 1.6)

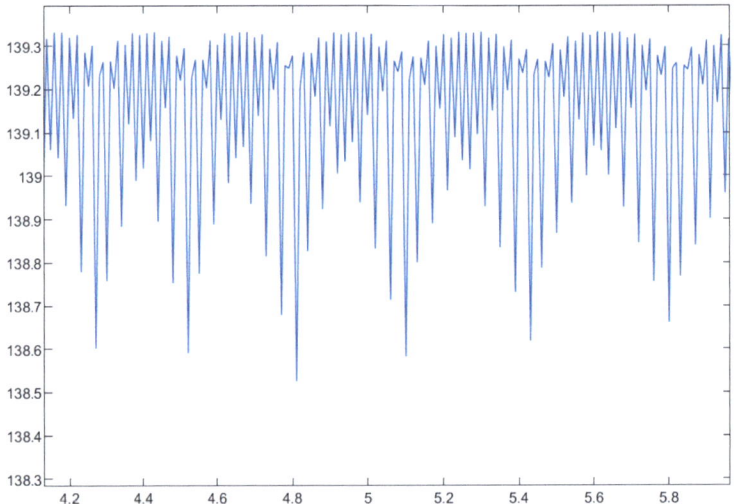

Fig. 1.22 A zoomed part of zone (**B**) of the obtained curve of temporal variation of the power provided by the used GPV (green box in Figs. 1.5 and 1.6)

1.5 Results and Discussions

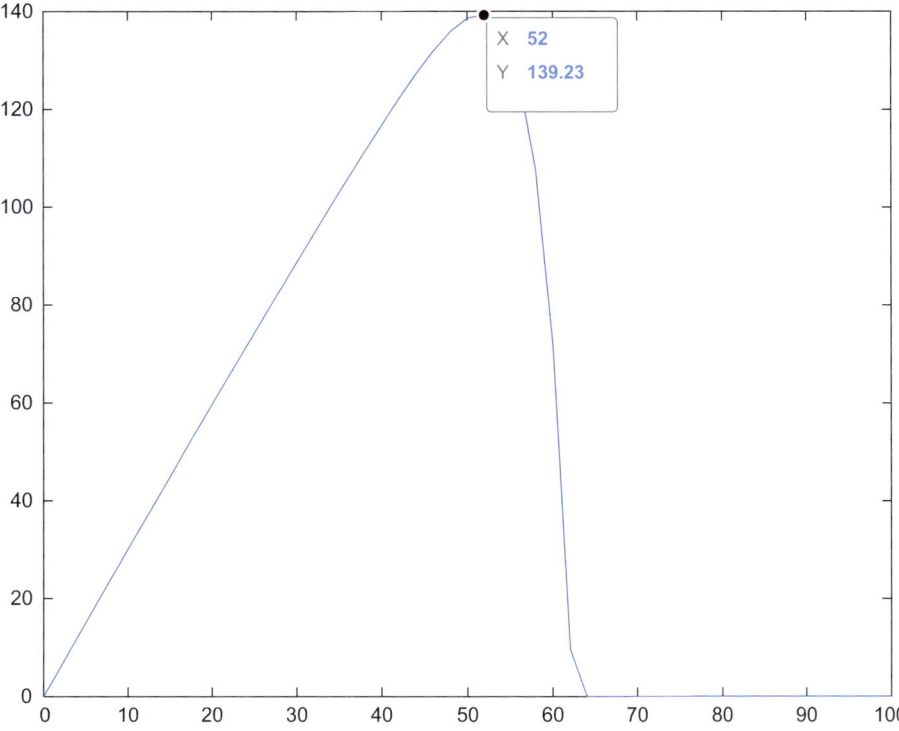

Fig. 1.23 The P–V characteristic obtained in case where the temperature, $T = 25\ °C$ and the Insolation, $G = 800\ W/m^2$

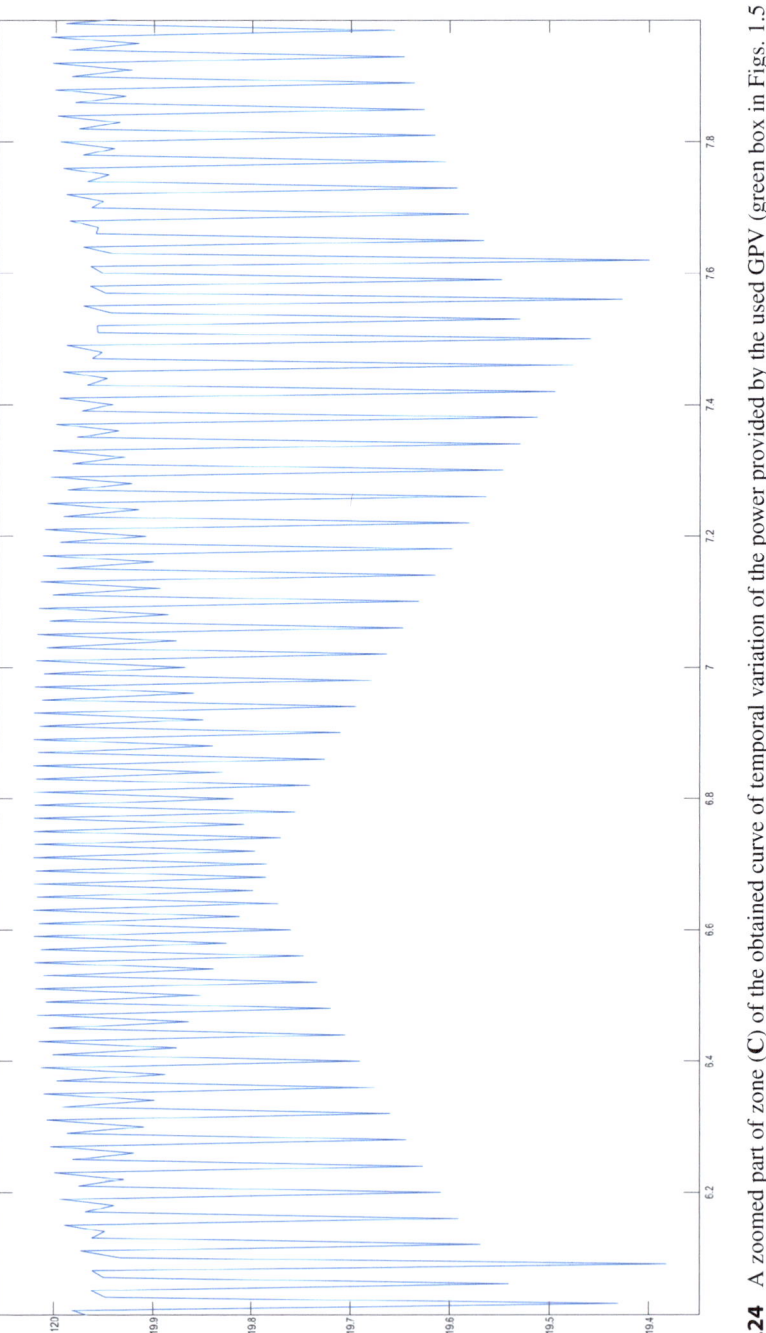

Fig. 1.24 A zoomed part of zone (C) of the obtained curve of temporal variation of the power provided by the used GPV (green box in Figs. 1.5 and 1.6)

1.5 Results and Discussions

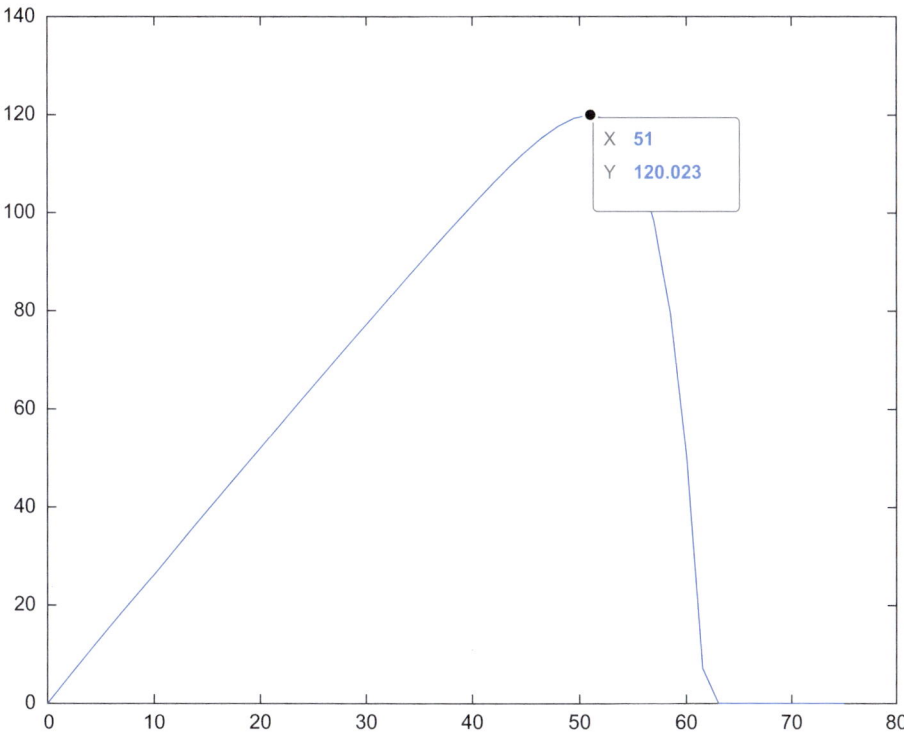

Fig. 1.25 The P–V characteristic obtained in case where the temperature, $T = 25\ °C$ and the Insolation, $G = 700\ W/m^2$

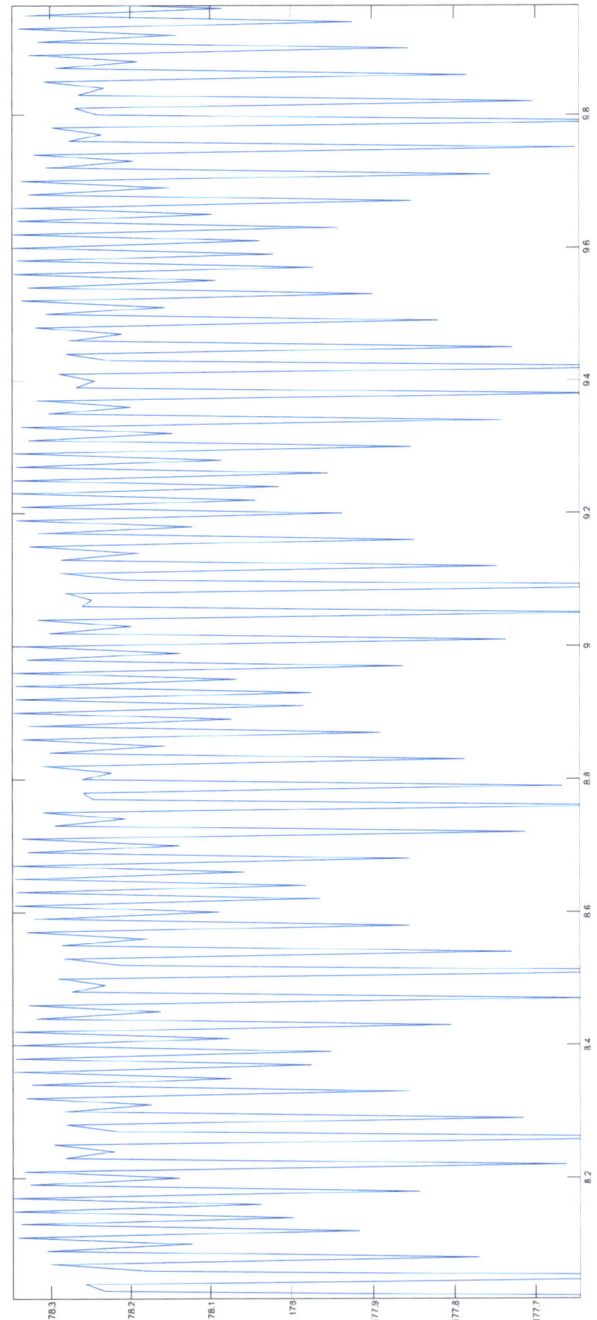

Fig. 1.26 A zoomed part of zone (**D**) of the obtained curve of temporal variation of the Power provided by the used GPV (green box in Figs. 1.5 and 1.6)

1.5 Results and Discussions

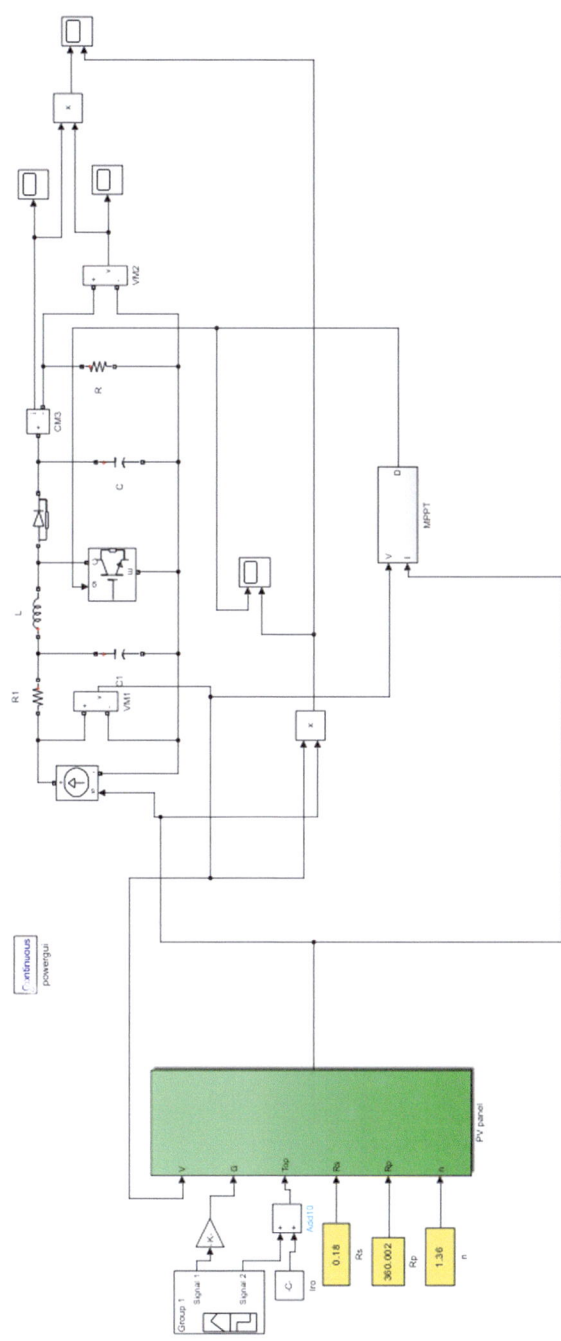

Fig. 1.27 The first PV system using IC controller with climatic conditions variables over time

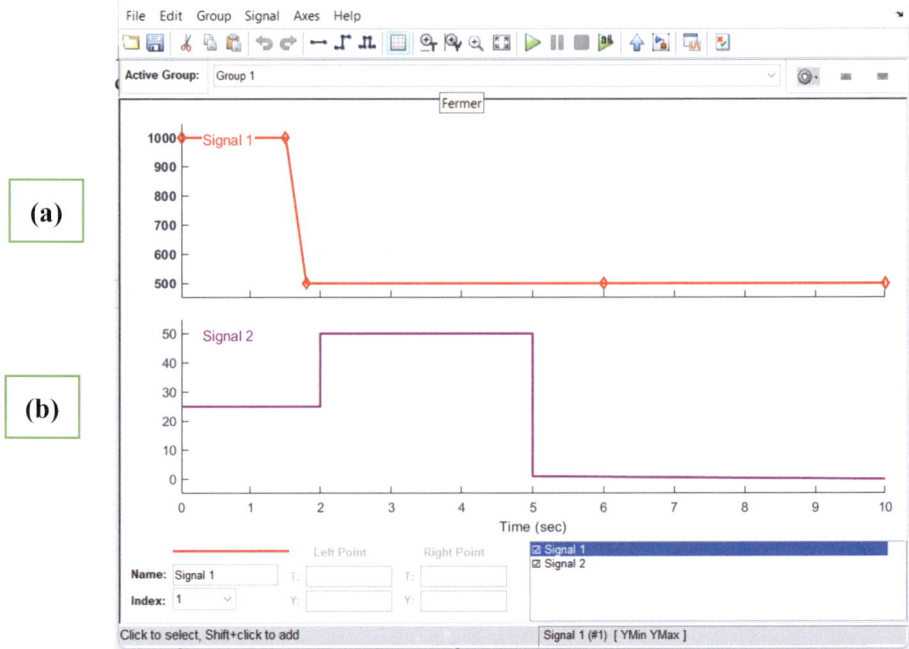

Fig. 1.28 a the curve of temporal variation of the insolation (Fig. 1.27), **b** the curve of temporal variation of the temperature (Fig. 1.27)

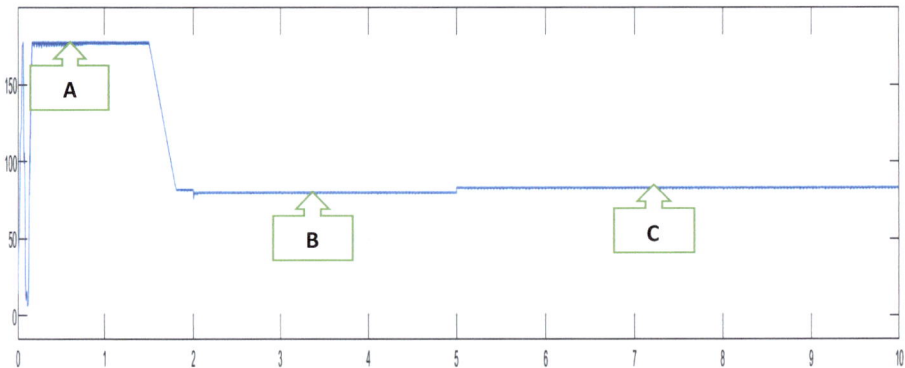

Fig. 1.29 The curve of temporal variation of the power provided by the used GPV (green box in Figs. 1.5 and 1.6), obtained in case where the climatic conditions are variables over time (Fig. 1.28)

1.5 Results and Discussions

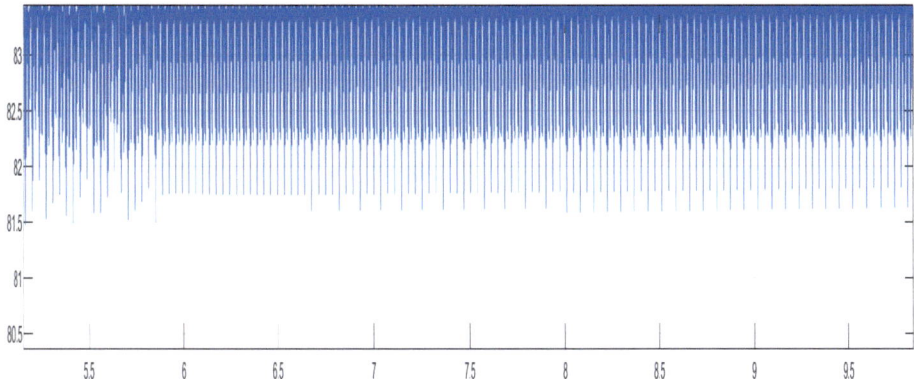

Fig. 1.30 A zoomed part of zone (**C**) of the obtained curve of temporal variation of the power provided by the used GPV (green box in Figs. 1.5 and 1.6)

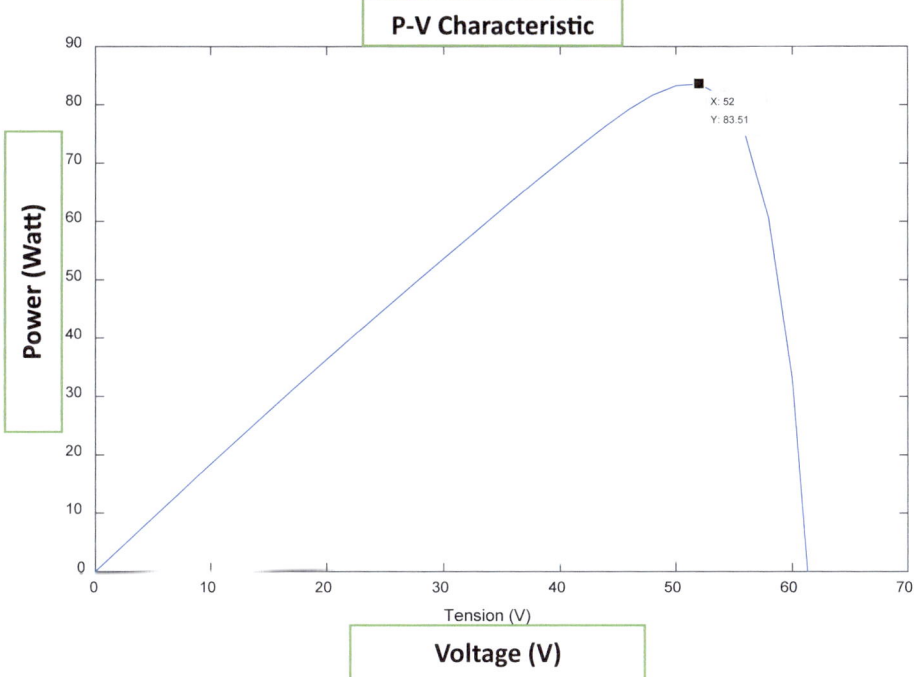

Fig. 1.31 P–V characteristic obtained in case where the insolation, $G = 500\ W/m^2$ and the temperature, $T = 0\ °C$

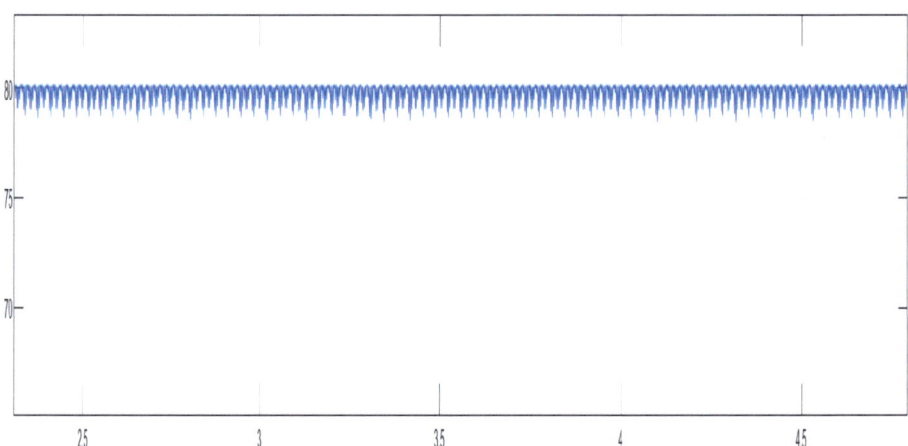

Fig. 1.32 A zoomed part of zone (**B**) of the obtained curve of temporal variation of the power provided by the used GPV (green box in Figs. 1.5 and 1.6)

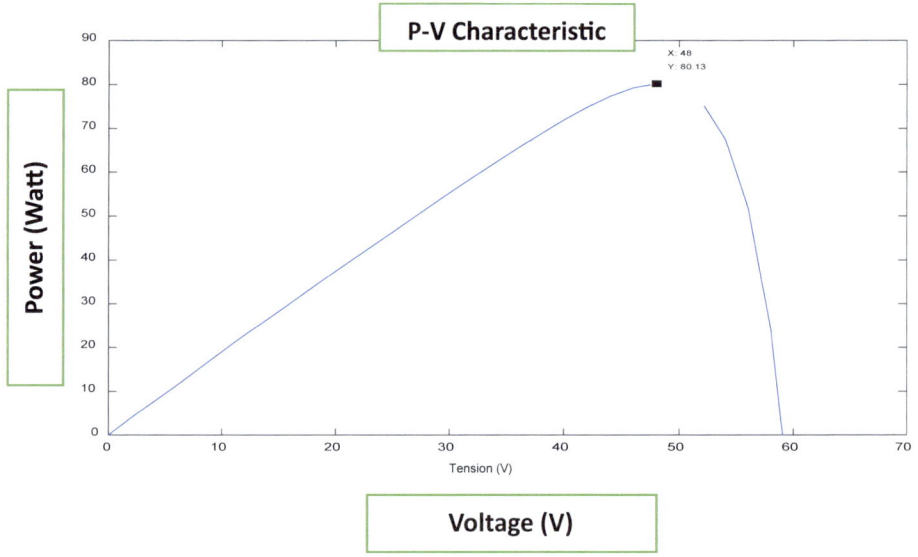

Fig. 1.33 P–V characteristic obtained in case where the insolation, $G = 500\ W/m^2$ and the temperature, $T = 50\ °C$

1.6 Conclusion

In this chapter, we are interested in two PV systems where the first includes a Perturb and Observe (P&O) controller and the second includes the Incremental Conductance (IC). The results obtained from the simulations of these PV systems show the performance of P&O and IC in tracking the MPP (Maximum PowerPoint) in the case of STC ($T = 25$ °C, $G = 1000\ W/m^2$) and in the case where the climatic conditions of temperature and insolation are variables over time.

References

1. H.D. Liu, C.H. Lin, K.J. Pai, Y.L. Lin, A novel photovoltaic system control strategies for improving hill climbing algorithm efficiencies in consideration of radian and load effect. Energy Convers. Manag. **165**, 815–826 (2018)
2. A.S. Mahdi, A.K. Mahamad, S. Saon, T. Tuwoso, H. Elmunsyah, S.W. Mudjanarko, Maximum power point tracking using perturb and observe, fuzzy logic and ANFIS. SN Appl. Sci. **2**, 89 (2020). https://doi.org/10.1007/s42452-019-1886-1
3. K.S. Tey, S. Mekhilef, Modified incremental conductance MPPT algorithm to mitigate inaccurate responses under fast changing solar irradiation level. Sol. Energy **101**, 333–342 (2014)
4. K. Ishaque, Z. Salam, H. Taheri, Modeling and simulation of photovoltaic (PV) system during partial shading based on a two-diode model. Simul. Model. Pract. TheoryPract. Theory **19**(7), 1613–1626 (2011)
5. R. Alik, A. Jusoh, T. Sutikno, A review on perturb and observe maximum power point tracking in photovoltaic system. Telecommun. Comput. Electron. Control **13**(3), 745 (2016)
6. H. Diab, H. El-Helw, H. Talaat, Intelligent Maximum Power Tracking and Inverter Hysteresis Current Control of Grid-Connected PV Systems, in *Proceedings of the International Conference on Advances in Power Conversion and Energy Technologies*, Mylavaram, pp.1–5 (2012)
7. H. Knopf, *Analysis, Simulation and evaluation of maximum power point tracking (MPPT) methods for a solar powered vehicle* (Portland state university, M.Sc, 1999)
8. T. Esram, P.L. Chapman, Comparison of photovoltaic array maximum power point tracking techniques. IEEE Trans. Energy Convers. **22**(2), 439–449 (2007)
9. A.I.M. Ali, Z.M. Alaas, M.A. Sayed, A. Almalaq, A. Farah, M.A. Mohamed, An efficient MPPT technique-based single-stage incremental conductance for integrated PV systems considering flyback central-type PV inverter. Sustainability (Switzerland) **14**, 1–15 (2022). https://doi.org/10.3390/su141912105
10. Z.S. Jalali, H.K. Hsien, N. Eskandarian, S. Mobayen, Improvement of self-predictive incremental conductance algorithm with the ability to detect dynamic conditions. Energies **14**(5), 1–14 (2021). https://doi.org/10.3390/en14051234
11. A. Asnil, R. Nazir, K. Krismadinata, M.N. Sonni, Performance analysis of an incremental conductance MPPT algorithm for photovoltaic systems under rapid irradiance changes. TEM J. **13**(2), 1087–1094, ISSN 2217–8309 (2024). https://doi.org/10.18421/TEM132-23
12. S. Amin, S. Khan, A. Qayoom (2018) Comparative analysis about the study of maximum power point tracking algorithm: a review. In: *International Mathematics (ICoMET). Conference on Computing, and Engineering Technologies*, pp. 1–8

13. M. Shixun, Y. Qintao, J. Kunping, M. Xiaofeng, S. Gengyu, An improved MPPT method for photovoltaic systems based on mayfly optimization algorithm. Energy Rep. **8**, 141–150 (2022). https://doi.org/10.1016/j.egyr.2022.02.160
14. MPPT based Photovoltaic (PV) system—File Exchange—MATLAB Central
15. A PHOTOVOLTAIC PANEL MODEL IN MATLAB/SIMULINK—File Exchange—MATLAB Central
16. Mppt (p and o)—File Exchange—MATLAB Central

Photovoltaic System's Modelling Based on Polytopic Transformation

2.1 Introduction

Nowadays, the issue of energy sources and the harmful increase of the greenhouse effect are more significant than ever. A pressing concern is the excessive consumption of natural resources, which reduces the reserves of such energy sources, threatening the earth's climate and consequently the well-being of future generations. Faced with this challenge, researchers are increasingly investigating new renewable energy sources while ensuring that these energies are non-polluting, inexhaustible, and sustainable. The primary renewable sources gaining significant attention in academia and industry are solar energy, biomass, geothermal, hydro, and wind energy [1]. The most commonly used renewable energy in both industry and daily life is solar energy. Solar power is exploited through photovoltaic cells that convert sunlight into electricity, making it widely accessible for residential use, commercial buildings, electric vehicles, water pumping, telecommunication, military space, and various industrial applications [2]. It's increasingly popular due to its affordability [3], Additionally, PV systems offer the advantage of being stationary making their installation, simple and quick compared to other renewable energy sources. As a result, they usually have a longer lifespan, often exceeding 20 years [4]. Despite the numerous advantages of PV energy, a major challenge still exists: the low efficiency of solar panels. This inefficiency results in greater space requirements for installation and higher overall costs.

The effective use of solar energy requires to track continuously the Photovoltaic panel's Maximum Power Point (MPP) which changes according to the climatic condition of illuminance, temperature, and wind speed. The accurate identification of MPP is closely tied to the controller's efficiency in managing this process. The most powerful and robust controllers nowadays are the intelligent controllers [5] whose development is based on the system model and their performance is linked to the model's accuracy. Thus, modeling is an important and decisive process for developing a performance controller and consequently maximizing photovoltaic power. Due to the complexity of the described system, these models are essential for predicting the I-V characteristics of the photovoltaic system, since they demonstrate non-linear characteristics as a function of solar intensity, angle of incidence, spectrum, and temperature [6, 7] Therefore, it is crucial to use a precise modeling approach to design PV systems, given the wide range of operating conditions and varying climates they must function in once installed.

This chapter proposes an accurate mathematical model for a PV system based on the multi-model approach. Using the convex polytopic transformation, the PV system can be represented by a combination of eight linear models derived from the system state equations. These models, called local models, are calculated according to the extreme values of temperature and insolation, in which, the benefit PV system might operate. The nonlinearity of the global system is rejected in eight weighting functions. Each weighting function is assigned to a local model. The PV system model, the global model, is generated by combining the local models assigned with the weighting functions. The obtained model is accurate, quite flexible, and able to describe the behavior of the PV system under various climatic conditions of temperature and insolation. The rest of this chapter is planned as follows: Sect. 2.2 describes the photovoltaic pumping system to be modeled. Section 2.3 gives the development of its model state. Section 2.4 details the polytopic transformation of the water-pumping system's state model and the elaboration of the multi-model. The simulation results of the proposed modeling method are given and discussed in Sect. 2.5. Finally, the conclusion and perspectives are provided in Sect. 2.6.

2.2 Photovoltaic Water Pumping System

Rural populations in the developing world face serious problems due to water shortages, particularly in arid regions. The lack of water in these areas poses a critical challenge for the residents. Consequently, searching for suitable solutions to this problem helps to improve the living conditions in these regions where the grid is unavailable. Photovoltaic solar pumping represents the perfect solution for water supply. In this chapter, we deal with photovoltaic pumping system modeling.

A photovoltaic water pumping system consists of a PV generator, a motor-driven pump, and a water reservoir. This setup ensures continuous water supply, even when solar energy is insufficient. Figure 2.1 illustrates the photovoltaic pumping system.

Two types of systems can be employed. In the first type, the Photovoltaic Generator (PVG) is directly connected to the pump, and the solar panels produce electricity that powers the DC pump. In the second type, a PV generator is connected to a converter (DC-AC or DC-DC) that helps in converting the generated DC electricity to AC or maintains the required DC voltage for driving a motor. We are concerned with the second type. We will model a photovoltaic system consisting of a PV generator coupled through a

2.2 Photovoltaic Water Pumping System

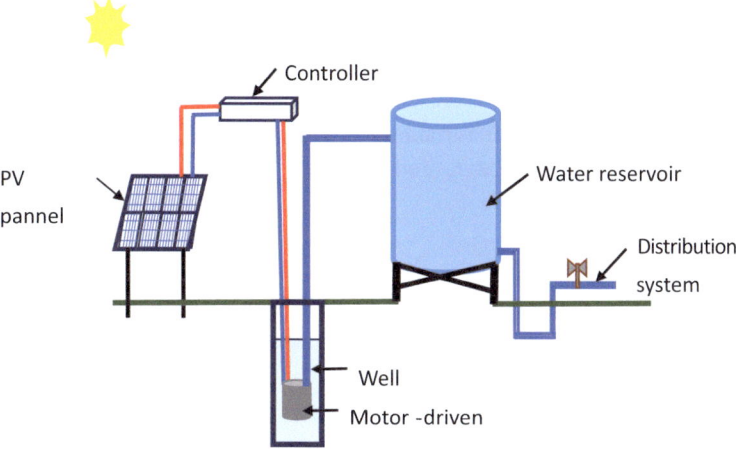

Fig. 2.1 Water pumping PV system

buck converter to a DC permanent magnet motor whose equivalent circuit diagram [2] is illustrated in Fig. 2.2.

It is made up of the following components:

- **Photovoltaic Generator (PVG)**

The PVG is constituted of N_p arrays that are connected in parallel. Each of these arrays is made up of n_p panels that are connected in series. Furthermore, each panel contains nc cells arranged in series. As a result, the total number of cells in the array is $N_s = n_p \times n_c$ cells. Consequently, when the cell current is denoted as I_{pvc} and its voltage as V_{pvc}, the PVG current and voltage are represented as $I_p = N_p I_{pvc}$ and $V_p = N_s V_{pvc}$, respectively.

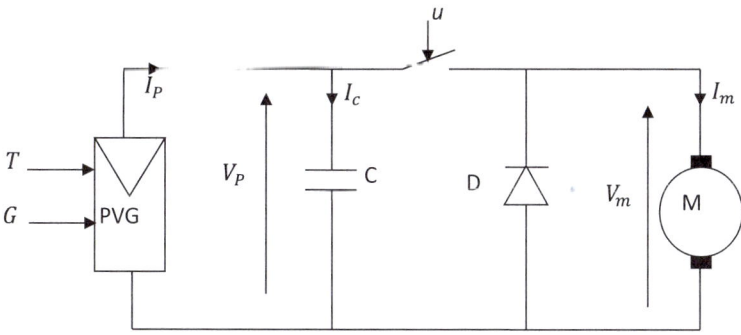

Fig. 2.2 Equivalent circuit diagram of the photovoltaic system

- **Buck converter**

The converter is connected to the capacitor C at the input, enabling the transformation of the current source I_p which represents the photovoltaic panel bank into a voltage source V_p. The switched control variable u commands the buck converter to allow the photovoltaic panel to generate the desired voltage.

- **The Diode**

The diode maintains a continuous flow of current within the motor at the moment when the power is disconnected, effectively preventing the occurrence of peak overvoltage.

- **Motor-driven pump**

We examine a direct current permanent magnet motor operating with constant flux, where the effect of armature reaction and switching phenomenon as supposed to be negligible. The operation of the motor pump is represented by Eqs. 2.1 and 2.2:

$$V_m = k_e \omega + L\frac{dI_m}{dt} + RI_m \quad (2.1)$$

$$J\frac{d\omega}{dt} = k_b I_m - (k_t + f)\omega \quad (2.2)$$

Table 2.1 summarizes the parameters of Eqs. 2.1 and 2.2.

It deserves mentioning that the current I_c in the capacitor C and the voltage V_m of the motor are expressed according to the switched control variable u as follows:

Table 2.1 Parameters of Eqs. 2.1 and 2.2

Parameter	Description
L	Electric inductance
k_e	Electromotive force constant
ω	Angular velocity
f	Motor viscous friction constant
k_t	Torque equation constant
J	Total inertia moment
k_b	Counter-electromotive force equation constant
R	Electric resistance
I_m	Current in the motor
V_m	Motor's voltage

$$I_c = I_P - u.I_m \qquad (2.3)$$

$$V_m = u.V_p \qquad (2.4)$$

2.3 State Model Elaboration

To elaborate a state model for the water-pumping PV system, we must adopt a mathematical representation for the PVG that simulates its energy output under varying environmental conditions. This representation can be deduced from the PV cell mathematical model.

The most generally adopted model for describing the complex voltage-current relationship in solar cells, defined by its non-linearity, is the single-diode model.

Due to its optimal balance of simplicity and precision, the single-diode model, given in Figure 2.3, is favored in practice. It outperforms the double-diode model in various features, such as enhanced accuracy in steady-state and fault diagnosis at the system level, the disposal of extensive datasets for diverse PV modules that are widespread in the marketplace, and the fast responsiveness within simulation settings [8–10]. In this model, the solar cell is presented by a single diode connected in parallel with a current source that characterizes the photovoltaic current. Additionally, it includes a series resistance that represents the cell's internal resistance to current flow, as well as a parallel resistance that accounts for leakage current [11–14].

The electrical behavior of a PV cell can be described, based on the single-diode model, by the following equation:

$$I_{pvc} = I_{phc} - I_{sd}\left(e^{q\frac{(V_{pvc}+R_S I_{Pvc})}{nKT_c}} - 1\right) - \frac{V_{pvc} + R_S I_{Pvc}}{R_{sh}} \qquad (2.5)$$

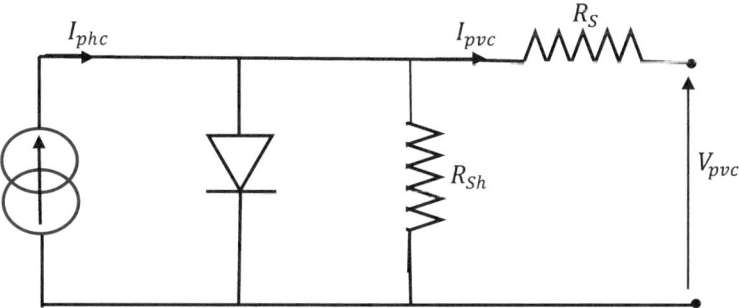

Fig. 2.3 Electrical representation of the single diode model

Table 2.2 Parameters of the PV cell equations

Parameter	Description
I_{pvc}	PV cell current
I_{phc}	Photocurrent
I_{sd}	Saturation current
V_{pvc}	PV cell voltage
R_S	Series resistance
R_{Sh}	Parallel resistance
n	Ideality factor of the diode
K	Boltzmann's constant $= 1.38 \times 10-23$
T	Cell temperature
q	Electron Charge $= 1.6 \times 10-19$

The parameters of the PV cell equation are summarized in Table 2.2.

In the photovoltaic generator, the output current I_P can be deduced according to the voltage V_P from Eq. (2.5) as follows:

$$I_p = N_p I_{phc} - N_p I_{sd} \left(e^{\frac{q}{nKT}\left(\frac{V_p}{N_s} + R_s \frac{I_p}{N_p}\right)} - 1 \right) - \frac{N_p V_p}{N_s R_{sh}} - \frac{R_s I_p}{R_{sh}} \quad (2.6)$$

The PV system's state equation is obtained by combining the Eqs. (2.1), (2.2), (2.3) and (2.4). It's given in Eq. (2.7).

$$\begin{cases} \frac{dV_P}{dt} = \frac{1}{C_e} I_P - \frac{1}{C_e} I_m.u \\ \frac{dI_m}{dt} = \frac{1}{L_m} V_P.u - \frac{R_m}{L_m} I_m - \frac{k_e}{L_m} \omega \\ \frac{d\omega}{dt} = \frac{k_b}{J} I_m - \frac{(k_t + F)}{J} \omega \end{cases} \quad (2.7)$$

Then, the operation of the PV system is described by three simultaneous equations, classifying it as a third-order system. Its state vector comprises three components. We defined the state vector $x = \begin{bmatrix} x_1 \\ x_2 \\ x_3 \end{bmatrix}$ where: $x_1 = V_p$; $x_2 = I_m$ and $x_3 = \omega$.

Then the state model of the water-pumping photovoltaic system is written as follows:

$$\begin{cases} \dot{x}_1 = \frac{1}{C_e} I_p - \frac{1}{C_e} x_2.u \\ \dot{x}_2 = \frac{1}{L_m} x_1.u - \frac{R}{L_m} x_2 - \frac{k_e}{L_m} x_3 \\ \dot{x}_3 = \frac{k_b}{J} x_2 - \frac{k_t + f}{J} x_3 \end{cases} \quad (2.8)$$

As exposed in Eq. (2.8), the complexity of the PV state model derives from I_p which vary non-linearly according to the voltage variable V_p. Therefore, if we note $I_p = h(x_1)$, the state model of the water-pumping photovoltaic system can be expressed as follows:

$$\begin{cases} \dot{x}_1 = \frac{1}{C_e}h(x_1) - \frac{1}{C_e}x_2.u \\ \dot{x}_2 = \frac{1}{L_m}x_1.u - \frac{R}{L_m}x_2 - \frac{k_e}{L_m}x_3 \\ \dot{x}_3 = \frac{k_b}{J}x_2 - \frac{k_t+f}{J}x_3 \end{cases} \quad (2.9)$$

The system's non-linearity arises from the exponential function present in the expression of I_p. This function prevents reliable linearization and complicates the independence between I_p and V_p. Consequently, this induces difficulties in command synthesis. In most cases, these challenges are addressed by approximating I_p around a fixed operating point [15], which negatively impacts the controller's precision and efficiency. In this chapter, we propose an efficient modeling method that does not require any simplification of the state model and guarantees the development of a highly accurate model.

2.4 Polytopic Transformation of the Water Pumping Photovoltaic System's State Model

2.4.1 Multi-model Modeling

One way to address the non-linear dynamic systems complexity problem is by reformulating the non-linear system to facilitate easier analysis. This is the principle of the Multi-model Modeling (MM) approach. This approach involves breaking down the complex dynamic behavior of the system into several operating zones, each defined by a local model. Depending on the specific zone, each model plays a different role in approximating the overall behavior of the complex system. Generally, the system displays consistent dynamic behavior within a specific operating zone. Consequently, the contribution of each local model to the global model, which is a convex combination of these models, is defined by a whitening function. Three main approaches are largely used in the literature to obtain a multi-model representation: identification, linearization, and transformation through non-linear sectors [16]. This chapter deals with the transformation through the non-linear sectors approach. This method involves a convex polytopic transformation of scalar function which is the origin of nonlinearity in the PV system state representation.

The class of linear model with Variable Parameters (LPV) systems, as introduced in [17] can be defined by a set of systems that admit a state representation in the following form:

$$\begin{cases} \dot{x} = A(\theta)x + B(\theta)u \\ y = C(\theta)x + D(\theta)u \end{cases} \quad (2.10)$$

where the parameter θ, which can be scalar or vectorial, varies over time, and generally involves within a bounded domain D_1. In many cases, the parameter of the LPV system coincides with either all or a portion of the initial state of the non-linear system. In such conditions, we will adopt a terminology widely used in the literature by referring to a quasi-linear model with Variable Parameters (quasi-LPV). Quasi-LPV modeling is exciting because it can accurately represent a vast class of non-linear systems. Through appropriate techniques, it becomes possible, for example, to evaluate stability. In particular, the photovoltaic pumping system discussed in this chapter can be represented using the quasi-LPV representation. For this purpose, we will divide h(x), contained in the state Eq. (2.9), by x_1. This division is possible since the state variable x_1 can vary with temperature and insolation between a non-zero minimum and a maximum values, therefore we can reformulate the system of Eq. (2.9) as follows:

$$\begin{cases} \dot{x}_1 = \frac{1}{C_e} h(x_1) \left(\frac{x_1}{x_1}\right) - \frac{1}{C_e} x_2.u \\ \dot{x}_2 = \frac{1}{L_m} x_1.u - \frac{R}{L_m} x_2 - \frac{k_e}{L_m} x_3 \\ \dot{x}_3 = \frac{k_b}{J} x_2 - \frac{k_t + f}{J} x_3 \end{cases} \quad (2.11)$$

The quasi-LPV form of the photovoltaic pumping system is then given by the Eq. (2.12).

$$\begin{bmatrix} \dot{x}_1 \\ \dot{x}_2 \\ \dot{x}_3 \end{bmatrix} = \begin{bmatrix} h(x_1)/x_1 & 0 & 0 \\ 0 & -R_m/L_m & -K_e/L_m \\ 0 & k_b/J & -(k_t+f)/J \end{bmatrix} \begin{bmatrix} x_1 \\ x_2 \\ x_3 \end{bmatrix} + \begin{bmatrix} -x_2/C_e \\ x_1/L_m \\ 0 \end{bmatrix} u \quad (2.12)$$

where:

$$\theta = [x_1, x_2, x_3] \quad (2.13)$$

$$A(\theta) = \begin{bmatrix} h(x_1)/x_1 & 0 & 0 \\ 0 & -R_m/L_m & -K_e/L_m \\ 0 & k_b/J & -(k_t+f)/J \end{bmatrix} \quad (2.14)$$

$$B(\theta) = \begin{bmatrix} -x_2/C_e \\ x_1/L_m \\ 0 \end{bmatrix} \quad (2.15)$$

2.4.2 Model's Base Establishment

The multi-model representation of the PV water pumping system will be derived from the state model representation given by Eq. (2.12). We define the set of non-constant variables in matrices A and B as a set of premises V_z [18]:

2.4 Polytopic Transformation of the Water Pumping Photovoltaic ...

where $V_z = \{z_1(x), z_2(x), z_3(x)\}$, with $z_1(x) = h(x_1)/x_1$; $z_2(x) = -x_2/C$; $z_3(x) = x_1/L$.

Each of these premises can be bounded by two extreme values that depend on temperature and irradiance.

$$\underline{z_1} < z_1(x) < \overline{z_1}; \ \underline{z_2} < z_2(x) < \overline{z_2}; \ \underline{z_3} < z_3(x) < \overline{z_3}$$

Since these premises are bounded, we can apply a polytopic transformation based on the following lemma.

Lemma 2.1 Any nonlinear function $f(x)$: $IR \rightarrow IR$.
Satisfying $\underline{f} < f < \overline{f} \forall x$.
Can be written as: $f(x) = f_1(x) \cdot \underline{f} + f_2(x) \cdot \overline{f}$.

where: $f_1 = \dfrac{\overline{f} - f(x)}{\overline{f} - \underline{f}}; f_2 = \dfrac{f(x) - \underline{f}}{\overline{f} - \underline{f}};$

The functions $f_1(x)$ and $f_2(x)$ satisfy the convex sum property: $f_1(x) + f_2(x) = 1$.

where:

$$0 < f_1(x) < 1 \text{ and } 0 < f_2(x) < 1 \forall x.$$

By applying the last lemma to the premises z_1, z_2, and z_3, we obtain:

$$z_1(x) = g_{11}(z_1(x))\underline{z_1} + g_{12}(z_1(x))\overline{z_1}$$

$$z_2(x) = g_{21}(z_2(x))\underline{z_2} + g_{22}(z_2(x))\overline{z_2}$$

$$z_3(x) = g_{31}(z_3(x))\underline{z_3} + g_{32}(z_3(x))\overline{z_3}$$

where:

$g_{11}(z_1(x)) = \dfrac{z_1(x) - \underline{z_1}}{\overline{z_1} - \underline{z_1}}$; $g_{12}(z_1(x)) = \dfrac{\overline{z_1} - z_1(x)}{\overline{z_1} - \underline{z_1}}$; and $g_{11}(z_1(x)) + g_{12}(z_1(x)) = 1$;

$g_{21}(z_2(x)) = \dfrac{z_2(x) - \underline{z_2}}{\overline{z_2} - \underline{z_2}}$; $g_{22}(z_2(x)) = \dfrac{\overline{z_2} - z_2(x)}{\overline{z_2} - \underline{z_2}}$; and $g_{21}(z_2(x)) + g_{22}(z_2(x)) = 1$;

$g_{31}(z_3(x)) = \dfrac{z_3(x) - \underline{z_3}}{\overline{z_3} - \underline{z_3}}$; $g_{32}(z_3(x)) = \dfrac{\overline{z_3} - z_3(x)}{\overline{z_3} - \underline{z_3}}$; and $g_{31}(z_3(x)) + g_{32}(z_3(x)) = 1$.

To simplify the writing, we will adopt the following notations:

g_{11}, and g_{12} instead of $g_{11}(z_1(x))$ and $g_{12}(z_1(x))$ respectively;

g_{21}, and g_{22} instead of $g_{21}(z_2(x))$ and $g_{22}(z_2(x))$ respectively;

g_{31}, and g_{32} instead of $g_{31}(z_3(x))$ and $g_{32}(z_3(x))$ respectively.

It is important to note that matrix $A(\theta)$ incorporates the premise z_1, while the matrix $B(\theta)$ involves the premises z_2 and z_3. Thus, we will estimate matrices A and B based on the vertex matrices of the polytope defined by the partition of the premises involved in these matrices. To achieve this, we begin with the expression of $A(\theta)$, which will be explicitly written in terms of z_1, as defined by the following.

$$A(\theta) = \begin{bmatrix} z_1 & 0 & 0 \\ 0 & -R_m/L_m & -K_e/L_m \\ 0 & k_b/J & -(k_t+f)/J \end{bmatrix} \quad (2.16)$$

The matrix A can be expressed as a summation of the matrices A_0 and A_1. Where:

$$A_0 = \begin{bmatrix} 0 & 0 & 0 \\ 0 & -R_m/L_m & -K_e/L_m \\ 0 & k_b/J & -(k_t+f)/J \end{bmatrix} \quad (2.17)$$

and

$$A_{z_1} = \begin{bmatrix} z_1 & 0 & 0 \\ 0 & 0 & 0 \\ 0 & 0 & 0 \end{bmatrix} \quad (2.18)$$

We can apply the polytopic transformation given in Lemma 2.1 to the matrix A_{z_1} which will be written as:

$$A_{z_1} = g_{11} \begin{bmatrix} \underline{z_1} & 0 & 0 \\ 0 & 0 & 0 \\ 0 & 0 & 0 \end{bmatrix} + g_{12} \begin{bmatrix} \overline{z_1} & 0 & 0 \\ 0 & 0 & 0 \\ 0 & 0 & 0 \end{bmatrix} \quad (2.19)$$

Then to reveal the partition functions g_{21}, g_{22}, g_{31} and g_{32}, we multiply A_{z_1} by the factor $(g_{21} + g_{22})(g_{31} + g_{32})$ which is equal to one. This results in the following expression of A_{z_1}:

2.4 Polytopic Transformation of the Water Pumping Photovoltaic ...

$$A_{z_1} = (g_{11}g_{21}g_{31} + g_{11}g_{21}g_{32} + g_{11}g_{22}g_{31} + g_{11}g_{22}g_{32}) \begin{bmatrix} z_1 & 0 & 0 \\ 0 & 0 & 0 \\ 0 & 0 & 0 \end{bmatrix}$$

$$+ (g_{12}g_{21}g_{31} + g_{12}g_{21}g_{32} + g_{12}g_{22}g_{31} + g_{12}g_{22}g_{32}) \begin{bmatrix} \overline{z_1} & 0 & 0 \\ 0 & 0 & 0 \\ 0 & 0 & 0 \end{bmatrix} \quad (2.20)$$

We notice:

$$A_1 = \begin{bmatrix} z_1 & 0 & 0 \\ 0 & 0 & 0 \\ 0 & 0 & 0 \end{bmatrix}; A_2 = \begin{bmatrix} \overline{z_1} & 0 & 0 \\ 0 & 0 & 0 \\ 0 & 0 & 0 \end{bmatrix}; \mu_1 = g_{11}g_{21}g_{31}; \mu_2 = g_{11}g_{21}g_{32};$$

$$\mu_3 = g_{11}g_{22}g_{31}; \mu_4 = g_{11}g_{22}g_{32}; \mu_5 = g_{12}g_{21}g_{31}; \mu_6 = g_{12}g_{21}g_{32};$$

$$\mu_7 = g_{12}g_{22}g_{31}; \mu_8 = g_{12}g_{22}g_{32}.$$

Then the state matrix $A(\theta)$, of the PV system is written as given in the Eq. (2.2. 2.21).

$$A(\theta) = A_0 + \mu_1 A_1 + \mu_2 A_1 + \mu_3 A_1 + \mu_4 A_1 + \mu_5 A_2 + \mu_6 A_2 + \mu_7 A_2 + \mu_8 A_2 \quad (2.21)$$

We apply the same method for the matrix B which can be written in terms of z_2, z_3 as follows:

$$B(\theta) = \begin{bmatrix} z_2 \\ z_3 \\ 0 \end{bmatrix} \quad (2.22)$$

We obtain the expression (2.23)

$$B(\theta) = g_{21} \begin{bmatrix} z_2 \\ 0 \\ 0 \end{bmatrix} + g_{22} \begin{bmatrix} \overline{z_2} \\ 0 \\ 0 \end{bmatrix} + g_{31} \begin{bmatrix} 0 \\ z_3 \\ 0 \end{bmatrix} + g_{32} \begin{bmatrix} 0 \\ \overline{z_3} \\ 0 \end{bmatrix} \quad (2.23)$$

To express the matrix $B(\theta)$ according of $\mu_i i = \{1, 2, 3, \ldots, 8\}$, we can proceed as follows:

- Multiply $g_{21} \begin{bmatrix} z_2 \\ 0 \\ 0 \end{bmatrix} + g_{22} \begin{bmatrix} \overline{z_2} \\ 0 \\ 0 \end{bmatrix}$ by the unit term $(g_{11} + g_{12})(g_{31} + g_{32})$

- Multiply $g_{31} \begin{bmatrix} 0 \\ z_3 \\ 0 \end{bmatrix} + g_{32} \begin{bmatrix} 0 \\ \overline{z_3} \\ 0 \end{bmatrix}$ by the unit term $(g_{11} + g_{12})(g_{21} + g_{22})$

- Denote by B_1; B_2; B_3 and B_4 the matrices $\begin{bmatrix} \overline{z_2} \\ \underline{z_3} \\ 0 \end{bmatrix}$; $\begin{bmatrix} \overline{z_2} \\ \underline{z_3} \\ 0 \end{bmatrix}$; $\begin{bmatrix} \overline{z_2} \\ \overline{z_3} \\ 0 \end{bmatrix}$ and $\begin{bmatrix} \overline{z_2} \\ \underline{z_3} \\ 0 \end{bmatrix}$ respectively.

Then $B(\theta)$ can be written according to the coefficients μ_i and the matrices B_j. $j = \{1, 2, 3, 4\}$ as the Eq. (2.24).

$$B(\theta) = (\mu_1 + \mu_5)B_1 + (\mu_2 + \mu_6)B_2 + (\mu_3 + \mu_7)B_3 + (\mu_4 + \mu_8)B_4 \quad (2.24)$$

As $\mu_1 + \mu_2 + \mu_3 + \mu_4 + \mu_5 + \mu_6 + \mu_7 + \mu_8 = 1$.
then: $(\mu_1 + \mu_2 + \mu_3 + \mu_4 + \mu_5 + \mu_6 + \mu_7 + \mu_8)A_0 = A_0$.
Consequently, the state equation of the photovoltaic water pumping system is expressed as follows:

$$\begin{aligned} A(\theta) + B(\theta) = &\ \mu_1(A_0 + A_1 + B_1 u) + \mu_2(A_0 + A_1 + B_2 u) + \mu_3(A_0 + A_1 + B_3 u) \\ &+ \mu_4(A_0 + A_1 + B_4 u) + \mu_5(A_0 + A_2 + B_1 u) \\ &+ \mu_6(A_0 + A_2 + B_2 u) + \mu_7(A_0 + A_2 + B_3 u) \\ &+ \mu_8(A_0 + A_2 + B_4 u) \end{aligned} \quad (2.25)$$

It is noted that the complex state model of the PV system is transformed into a weighted summation of linear subsystems without using any simplification or the need of functioning point identification. This approach simplifies the system analysis and facilitates the easy computation of a performance controller.

2.5 Results and Discussions

In this section, we will examine a photovoltaic pumping system with the characteristics listed in Table 2.3.

To validate the reliability of the developed model, we simulate its outputs (state vector components) under varying temperature and illuminance conditions and compare them with the PV system's outputs under identical conditions. Figure 2.4 illustrates the evolution of the Photovoltaic generator voltage V_p and the panel voltage V_{pm} generated by the developed model while considering a temperature range from 0 °C to 70 °C and an illuminance range from 200 to 1000 W/m².

We observe that the curves presented in Fig. 2.4 appear identical. To further ensure the model precision, it's essential to represent the error between the Photovoltaic generator voltage V_p and the panel voltage V_{pm} under varying climatic conditions of temperature and illuminance. Figure 2.5 illustrates the progression of this error across a range of

2.5 Results and Discussions

Table 2.3 Characteristics of the PV system

	Parameters	Values
Photovoltaic generator	V_{oc}	36.6
	I_{Sc}	8.28
	V_{mp}	32
	I_{mp}	7.6
	N_S	60
	α_0	0.05%
	β_0	−0.34% °C
Moto-pump	Nominal voltage u_n	22 V
	Nominal current I_n	10 A
	Rotational speed ω_n	1500 tr/mn
	R	1 Ω
	L	0.01H
	f	$880 \cdot 10^{-6}$ N.m.$\frac{S}{rd}$
	k_e	0.5
	k_t	0.1
	J	$470 \cdot 10^{-6}$ kg.m^2
	C	$47 \cdot 10^{-4}$ F
Simulation parameters	Maximum temperature	70 °C
	Minimum temperature	0 °C
	Maximum illuminance	1000 W/m^2
	Minimum illuminance	200 W/m^2

temperatures from 0 to 70 °C and a range of illuminances from 200 W/m^2 to 1000 W/m^2.

The results shown in the last figure indicate a very weak error of 9.2310^{-14}, which proves that the model has successfully reproduced the same voltage of the PV generator under all given climatic conditions of temperature and illuminance.

To further ensure the high accuracy of this model, we will evaluate its ability to replicate the same motor current I_m and rotational speed ω_m of the real system. Therefore, the error between the motor current I_m and the model's current I_{mm} and that between the motor rotational speed ω and the model's rotational speed ω_m are assessed across a range of temperatures from 0 to 70 °C and a range of illuminances from 200 W/m^2 to 1000 W/m^2. The evolution of the error ($I_m - I_{mm}$) is presented in Fig. 2.6 and the evolution of the error ($\omega - \omega_m$) is presented in Fig. 2.7.

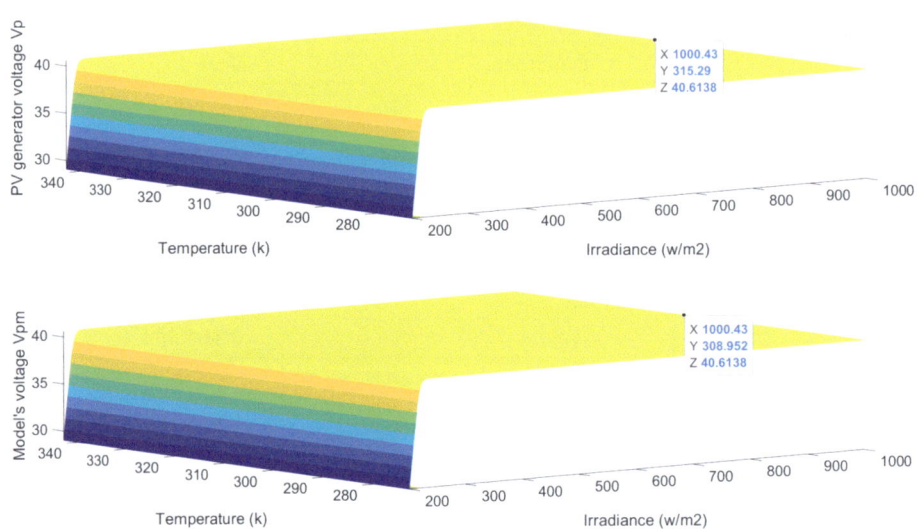

Fig. 2.4 Evolution of the Photovoltaic generator voltage V_p and the panel voltage V_{pm} for temperature varying from 0 to 70 °C and illuminances varying from 200 W/m² to 1000 W/m²

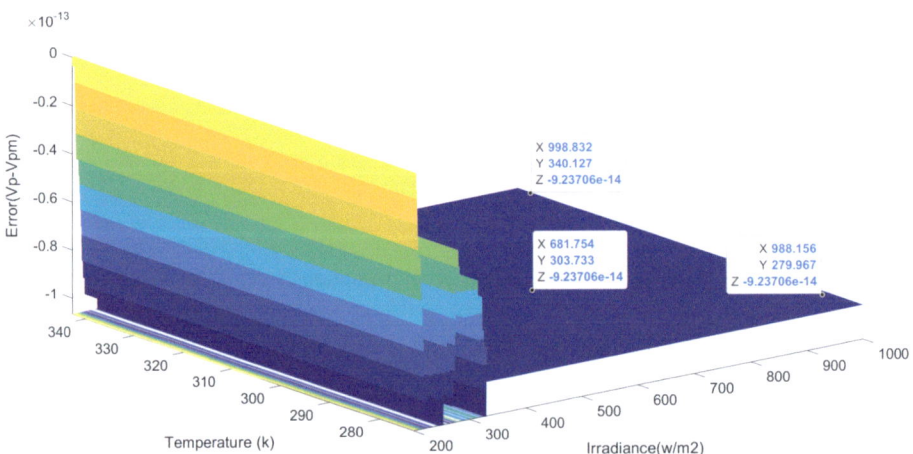

Fig. 2.5 Progression of the error between the panel voltage and the model's voltage V_{pm} across a range of temperatures from 0 °C to 70 °C and illuminances from 200 W/m² to 1000 W/m²

2.5 Results and Discussions

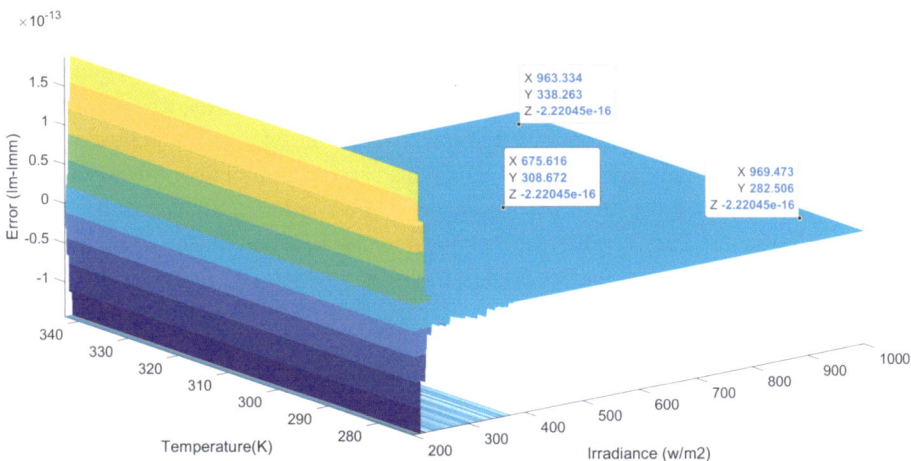

Fig. 2.6 Evolution of the error between the motor current I_m and the model's current I_{mm} across a range of temperatures from 0 °C to 70 °C and illuminances from 200 W/m² to 1000 W/m²

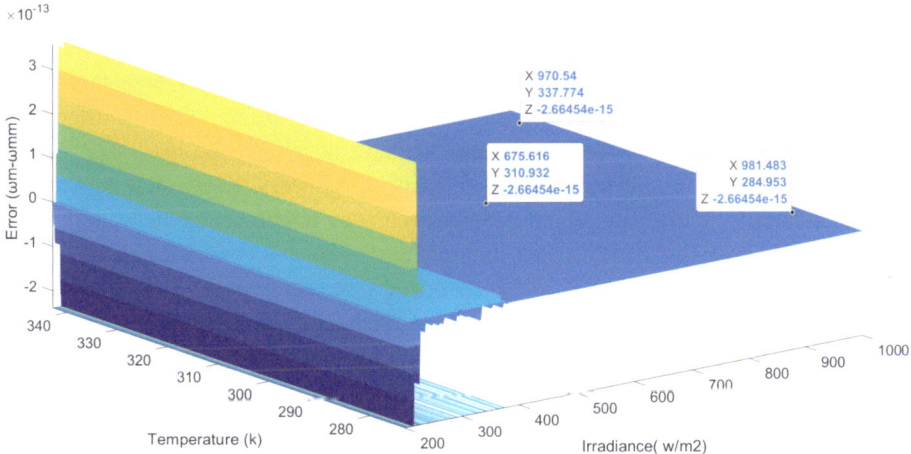

Fig. 2.7 Evolution of the error between the motor rotational speed ω and the model's rotational speed ω_m across a range of temperatures from 0 °C to 70 °C and illuminances from 200 W/m² to 1000 W/m²

Figures 2.6 and 2.7 depict a current modeling error of 2.22 10^{-16} and a motor rotational speed error of 2.66 10^{-15} respectively. These very weak errors indicate extremely low error margins. Such low inaccuracy levels reflect the high quality of the generated model, as well as its capacity to reproduce the PV system's behavior very accurately. The reliability and accuracy demonstrated by the model across this wide range of climatic conditions emphasize its dependability and effectiveness in real-world applications. This level of accuracy not only accentuates the model's capability to deliver precise representation of the complex PV system but also underlines its potential for larger deployment in other PV systems. Such small error margins demonstrate the efficacy of the modeling efforts and the advanced approaches used in their development.

2.6 Conclusion

In this chapter, we have introduced a thoughtfully designed model structure aimed at accurately reflecting the behavior of a Photovoltaic pumping system and its expected variations in performance due to temperature and solar radiation changes. The development of the precise model that mirrors the system's behavior is achieved by employing a set of linear and simple models, each is computed based on extreme values of temperature and solar radiation to which the PV system can be exposed. The modeling process begins with transforming the state equation of the PV system into the quasi-linear parameter varying form, which is crucial for accommodating the varying parameters of the system. We then identified the premise variables that significantly influence the model's performance. These variables are carefully decomposed, according to their extremum, to facilitate a detailed analysis. Following this, we have developed the necessary matrices and weighting functions fundamental to the modeling process. Every step is performed with the greatest care to ensure that the model accurately represents the delicate functionality of the PV pumping system under different environmental conditions of temperature and illumination. In the last step of the modeling procedure, we created eight specific linear models assigned with nonlinear weighted functions where the nonlinearity of the PV system is rejected. The simplicity and efficiency of these partial models facilitate the implementation of a robust and real-time control law. Each partial model allows the generation of a partial controller. These partial controllers are then combined to form a comprehensive global control strategy for the PV system. The accuracy and reliability of the established model are apparent through the simulation results presented in this chapter, assisting in establishing the potential for real-world applications. Because it is based on integrating multiple simple models, our approach is easily adaptable to other photovoltaic systems. Such development promotes the advancement of new technological advances, as well as their use in the development of renewable energy resources. To summarize, we have discussed a type of PV pumping systems and demonstrated a dependable and adaptable model that best fits the parameters of this specific device. This role introduces new ideas and directions in energy, as well as encourages additional research and innovative approaches for improving the effectiveness and dependability of renewable energy systems.

References

1. A.V. Herzog, T.E. Lipman, D.M. Kammen, Renewable Energy Sources. University of California, Berkeley (2001)
2. M. Nawel, T. Mourad, B. Mongi, Modelling of photovoltaic water pumping system based on artificial intelligence. Adv. Model. Anal. B. **62**(1), 11–17 (2019)
3. A.S. Al-Ezzi, M.N.M. Ansari, Photovoltaic solar cells: a review. Appl. Syst. Innov. **5**(4) (2022)
4. C. Bucher, J. Wandel, D. Joss, Live expectancy of PV inverters and optimizers in residential PV systems, in *8th World Conference on Photovoltaic Energy Conversion (WCPEC-8)* (2022)
5. A. Baciu, C. Lazar, Comparative performance analysis of intelligent PID controllers for a mechatronic system. Bull. Polytech. Inst. Iași. Electr. Eng. Power Eng. Electron. Sect. **69**(3) (2023)
6. E.A. Silva, F. Bradaschia, M.C. Cavalcanti, A.J. Nascimento, Parameter estimation method to improve the accuracy of photovoltaic electrical model. IEEE J. Photovolt. **6**, 278–285 (2015)
7. E.A. Silva, F. Bradaschia, M.C. Cavalcanti, A.J. Nascimento, Renewable energy-utilisation and system integration. IEEE J. Photovolt. **6**, 278–285 (2006)
8. M. Suthar, G.K. Singh, R.P. Saini, Comparison of mathematical models of photo-voltaic (PV) module and effect of various parameters on its performance, in *Proceedings of the 2013 International Conference on Energy Efficient Technologies for Sustainability (ICEETS)* (Nagercoil, India, 2013), pp. 1354–1359
9. V. Quaschning, R. Hanitsch, Numerical simulation of current-voltage characteristics of photovoltaic systems with shaded solar cells. Sol. Energy **56**, 513–520 (1996)
10. A.N. Kumar, M. Sameer, M.M. Rasheed, A. Balakrishna, S.N. Venkat, V.A Manasa, Detailed modeling and simulation of photovoltaic module, in *Proceedings of the 2020 International Conference on Smart Technologies in Computing, Electrical and Electronics (ICSTCEE)* (Bengaluru, India, 2020), pp. 39–43
11. K.I. Baradieh, M.A.A.M. Zainuri, N.A. Mohamed Kamari, H. Abdullah, Y. Yusof, M.A. Zulkifley, M.A. Koondhar, A study on the impact of different PV model parameters and various DC faults on the characteristics and performance of the photovoltaic arrays **9**(5), 93 (2024)
12. H.J. El-Khozondar, R.J. El-Khozondar, K. Matter, Parameters influence on MPP value of the photo-voltaic cell. Energy Procedia **74**, 1142–1149 (2015)
13. V. Stornelli, M. Muttillo, T. de Rubeis, I. Nardi, A new simplified five-parameter estimation method for single-diode model of photovoltaic panels. Energies **12**(22), 4271 (2019)
14. M.A. Tofael, G. Teresa, Fault analysis in solar photovoltaic arrays, in *Proceedings of the 5th International Conference on Renewable Energy Research and Applications* (Bermigham, UK, 2016), pp. 20–23
15. D. Abbes, C.H. Gerard, M. Andre, R. Benoit, Modeling and simulation of a photovoltaic system: an advanced synthetic study, in *3rd International Conference on Systems and Control*, pp. 29–31 (2013)
16. E. Nesrine, Approche neuronale de la representation et de la commande multimodèle de processus complexe. Ph.D. Thesis. Ecole Doctorale Science pour l'Ingénieur Université Lille, France (2010)

17. B. Jean-Marc, Commande robuste des systèmes à paramètres variables', Ph.D. Thesis. L'Ecole Nationale Supérieure de l'Aéronautique et de l'Espace , France (1996)
18. M.N. Anca, Analyse et synthèse de multimodèle pour le diagnostic. Application à une station d'épuration'. Ph.D. Thesis. National Polytechnic Institute of Lorraine, France (2010)

Effective PI Controller for Maximum Power Point Tracking of a Photovoltaic System

3.1 Introduction

While solar energy has numerous advantages such as renewability, sustainability, wide availability, and environmental friendliness, it is still constrained by the nature of photovoltaic (PV) panels [1], which have relatively low productivity compared to other sources of energy. Fortunately, recent technological developments have resulted in enhancements [2]. For instance, cutting-edge monocrystalline panels utilizing N-type cells are now capable of surpassing efficiencies of 24%. The latest silicon-based monocrystalline panels can even achieve performance levels of 24.2% with industry leaders like Aiko Solar at the forefront with their Neostar Series panels [3]. Novel technologies, like perovskites, are gaining attention for their high-efficiency levels of 25% establishing them as players in the renewable energy industry landscape of options for power generation and storage systems, in the future and beyond. Nonetheless, considerable effort remains necessary to enhance the overall efficiency and output of photovoltaic panels [1, 4]. Researchers have been motivated by this need for improvement, as they are focused on PV systems with higher energy conversion efficiency. Photovoltaic (PV) systems work most effectively, producing peak power output, at the Maximum Power Point (MPP). The MPP is a special point on both, the current–voltage (I-V) as well as the power-voltage (P-V) charactcristic curves of the PV system. The MPP is extremely sensitive and easily altered by temperature and illuminance variations [5]. Factors like varying sunlight, shading from clouds or surrounding structures, and temperature changes can all lead to an MPP shift. As a result, in order to run optimally, the system must adhere to the MPP consistently. Maximum PowerPoint Tracking (MPPT) systems that utilize search algorithms gain popularity through their straightforward design and affordability. The traditional MPPT systems suffer from oscillations close to the MPP [6] and decreased performance under changing temperature and illuminance conditions [7]. Researchers have investigated advanced methods including Artificial Neural Networks (ANN) [6] genetic algorithms [8], and fuzzy logic [9] to address these problems. These advanced techniques effectively handle parameter variations but require substantial computational resources [10]. Their requirement for substantial computational resources leads to reduced practicality for routine application. PV systems continue to require an economical and efficient controller with robust capabilities [11]. PV applications have relied on proportional-integral (PI) controllers for many years [12]. A DC-DC converter with a PI controller enables the PV system to adjust its nominal voltage for maximum power delivery. These controllers deliver reliable performance through their simple design. The state-space representation of the DC/DC converter involves state variables that depend on the control variable which makes PI controller synthesis via transfer functions challenging. Linearization is typically performed around the intended operating point to approximate the transfer function [13]. However, the MPP fluctuates across the operation domain, preventing stable control via linearization at a single fixed point. The design should enhance PI controller performance for MPPT while keeping its simplicity intact. The literature widely adopts gain tuning of controllers that happens throughout the operational period [14, 15, 16]. Online adaptation methods which use optimization techniques are commonly required for this approach but these methods lead to slowed Maximum Power Point achievement and higher implementation expenses. System modeling is the first engineering stage. It serves as the foundation for comprehending and investigating complicated systems. When dealing with complex and highly non-linear processes, a multi-model structure is a useful way to address controller design issues. The suggested approach is based on substituting a series of simplified models that reflect the behavior of the system in a specific operating domain. This strategy enables engineers to easily manage and control processes in a range of disciplines, including chemistry, biology, and industry, by creating a multi-model control approach based on this framework [17, 18, 19].

This study adopts a multi-model perspective to develop a new synthesis technique for PI controllers in photovoltaic systems. This approach intends to keep the traditional PI controllers' merits and eliminate their intrinsic disadvantages. More specifically, the method which is based on the polytopic transformation is applied to convert the complex PV system state model, comprising a Photovoltaic Generator (PVG) and a buck converter, into a combination of eight linear models assigned specific weighting functions. A partial PI controller is designed to follow a desired reference trajectory for each partial model. The partial controller gains are carefully tuned to ensure rapid and precise maximum power point tracking while maintaining stability within two linear control loops: the inductor current control loop and the control loop of a specified voltage function. Each partial controller's proportional and integral gains are weighted using the nonlinear weighting function associated with the partial model under control. The weighted gains are then combined to calculate the proportional and integral gains of the PV system's global PI controller.

3.2 Representation of the PV System State Model

The chapter is organized as follows: In Sect. 3.2, we provide a thorough description of the PV system and its state model. The polytopic transformation of the PV system is detailed in Sect. 3.3. Section 3.4 elaborates further on the formulation of the partial controllers for linear models and the design of the global PI controller. Simulation results for the different irradiance and temperature profiles are shown and discussed in Sect. 3.5, as well as a comparison between the proposed method and the traditional PI controller. Lastly, the chapter is concluded in Sect. 3.6.

3.2 Representation of the PV System State Model

The Maximum PowerPoint of a solar panel is significantly affected by temperature and irradiance variations. These variations create different optimal operating points under different environmental conditions. The direct coupling of a Photovoltaic generator to its load does not guarantee operation at the MPP at all times because the system is incapable of autonomously compensating for such variations. For instance, a rise of irradiance increases the current produced by the solar panel and slightly its voltage, leading to greater power output. Equally, temperature increases mostly decline the voltage output. This results in reduced overall power.

To address this issue, a DC/DC converter is employed between the PVG and the load. The system manages the duty cycle of the converter to maintain a voltage and current matching the MPP despite temperature or irradiance changes [20]. Maximum PowerPoint Tracking algorithms monitor power output and modify the converter duty cycle accordingly. Using a DC/DC converter with MPPT allows a PV system to respond to its surroundings in real-time, resulting in high energy efficiency across a wide range of climatic conditions. In this chapter, we will deal with a photovoltaic system controlled through a buck converter. The circuit diagram of this system is given in Fig. 3.1.

In this circuit, the buck converter is necessary to reduce the voltage generated by the photovoltaic generator (PVG) to a suitable level for the connected load. The device consists of the inductor (L), responsible for regulating the current flow and smoothing the fluctuations in the output current I_L, it is vital to identify the buck converter's proper operation and capability, the capacitor C_e responsible for filtering and stabilizing the

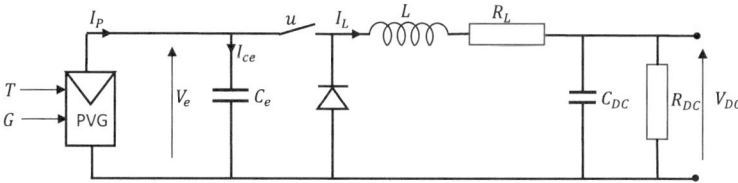

Fig. 3.1 Circuit diagram of the studied system

voltage V_e by reducing ripple and noise. The capacitor C_{DC} plays the same role as the capacitor C_e regarding the voltage V_{DC} to stabilize a constant DC supply to the load. The diode serves as a safety to ensure that the current flows naturally by allowing it to flow in one direction and avoiding backflow. The Resistor R_{DC} represents the load supplied by the electrical energy from the circuit. It may be any electrical device that requires a specific voltage and current to operate. The switch (typically a transistor or MOSFET) which is controlled by the control variable u manipulates the energy transfer from the PVG to the load by regulating the output voltage by turning on and off. On the other hand, the photovoltaic generator represents the main power source of the system, converting the sunlight through the photovoltaic effect to direct current (DC) electricity.

The following simultaneous equations define the operation of this circuit:

$$\begin{cases} \frac{dV_P}{dt} = \frac{1}{C_e}I_P - \frac{1}{C_e}I_L.u \\ \frac{dI_L}{dt} = \frac{1}{L}V_P.u - \frac{R_L}{L}I_L - \frac{1}{L}V_{DC} \\ \frac{dV_{DC}}{dt} = \frac{1}{C_{DC}}I_L - \frac{1}{R_{DC}C_{DC}}V_{DC} \end{cases} \quad (3.1)$$

where:

$$I_p = I_{ph} - I_s \left(e^{q\frac{(V_p+R_S I_p)}{nN_S KT_c}} - 1 \right) - \frac{V_p + R_S I_p}{R_{sh}} \quad (3.2)$$

The parameters of the Eq. (3.2) are summarized in Table 3.1.

It's a third-order system with a state vector X comprising three components x_1, x_2 and x_3. We consider: $x_1 = V_P$, $x_2 = I_L$, and $x_3 = V_{DC}$. The equation system (3.1) is expressed according to the vector X as stated below:

Table 3.1 parameters of the PV cell equations

Parameter	Description
I_p	PVG current
I_{ph}	Photocurrent
I_s	Saturation current
V_p	PVG voltage
R_S	Series resistance
R_{Sh}	Parallel resistance
n	Ideality factor of the diode
K	Boltzmann's constant $= 1.38 \times 10-23$
T	Cell temperature
q	Electron charge $= 1.6 \times 10-19$

3.3 PV System State Model Transformation Using Polytopic ...

$$\begin{cases} \dot{x}_1 = \frac{1}{C_e} I_P - \frac{1}{C_e} x_2 . u \\ \dot{x}_2 = \frac{1}{L} x_1 . u - \frac{R_L}{L} x_2 - \frac{1}{L} x_3 \\ \dot{x}_3 = \frac{1}{C_{DC}} x_2 - \frac{1}{R_{DC} C_{DC}} x_3 \end{cases} \quad (3.3)$$

Since the current I_P provided by the photovoltaic generator depends on the voltage V_P, as revealed in Eq. (3.2), we can rewrite the last equation as follows:

$$\begin{cases} \dot{x}_1 = \frac{1}{C_e} \theta(x_1) - \frac{1}{C_e} x_2 . u \\ \dot{x}_2 = \frac{1}{L} x_1 . u - \frac{R_L}{L} x_2 - \frac{1}{L} x_3 \\ \dot{x}_3 = \frac{1}{C_{DC}} x_2 - \frac{1}{R_{DC} C_{DC}} x_3 \end{cases} \quad (3.4)$$

It's evident that to develop a PI controller for controlling the switch, it's essential to establish the transfer function from the state Eq. (3.4). However, as the control variable u affects the state components x_1 and x_2 within this equation, it becomes challenging to accomplish the task. Researchers address this challenge by linearizing the state equation around an equilibrium point characterized by specific temperature and illuminance [13]. Nonetheless, this can easily lead to system destabilization when the climatic conditions experience a noticeable change. An innovative approach is introduced in this chapter for synthesizing an efficient PI controller without relying on linearization around a single operating point. This approach is based on the multi-model structure elaborated for the PV systems by Mensia et al. [21].

3.3 PV System State Model Transformation Using Polytopic Transformation

Our objective is to transform the complex state representation of the PV system into a simplified representation that facilitates the synthesis of an efficient and robust PI controller. To achieve this, we use the method proposed and detailed in Chap. 2, rewriting the PV system state equation as the following quasi-linear model with variable parameters.

$$\dot{X} = A(\delta(X, u))X + B(\delta(X, u))u \quad (3.5)$$

where:

$$A(\delta(X, u)) = \begin{bmatrix} \theta(x_1)/x_1 & 0 & 0 \\ 0 & -R_L/L & -1/L \\ 0 & 1/C_{DC} & -1/R_{DC} C_{DC} \end{bmatrix} \quad (3.6)$$

$$B(\delta(X, u)) = \begin{bmatrix} -x_2/C_e \\ x_1/L \\ 0 \end{bmatrix} \quad (3.7)$$

It is worth mentioning that $\delta(X, u)$ in matrix A is different from $\delta(X, u)$ in matrix B. For simplicity, however, we use the same notation in both matrices A and B. It's also important to note that the division by x_1 in matrix A is allowed because x_1 can only vary within a strictly positive range of values. The premises should be the non-constant parameters existing in matrices A and B. Therefore, we obtain three premises: $z_1(x) = \theta(x_1)/x_1$; $z_2(x) = -x_2/C_e$; and $z_3(x) = x_1/L$.

As the premises are bounded, they can be expressed according to their extremums by the following formulas:

$$z_1(x) = f_{11}\underline{z_1} + f_{12}\overline{z_1} \tag{3.8}$$

$$z_2(x) = f_{21}\underline{z_2} + f_{22}\overline{z_2} \tag{3.9}$$

$$z_3(x) = f_{31}\underline{z_3} + f_{32}\overline{z_3} \tag{3.10}$$

where:

$\overline{z_1}$, $\overline{z_2}$, and $\overline{z_3}$ are the maximum of z_1, z_2, and z_3 respectively;
$\underline{z_1}$, $\underline{z_2}$, and $\underline{z_3}$ are the minimum of z_1, z_2, and z_3 respectively;

$$f_{11} = \frac{z_1(x) - \underline{z_1}}{\overline{z_1} - \underline{z_1}}; f_{12} = \frac{\overline{z_1} - z_1(x)}{\overline{z_1} - \underline{z_1}}; f_{11} + f_{22} = 1 \tag{3.11}$$

$$f_{21} = \frac{z_2(x) - \underline{z_2}}{\overline{z_2} - \underline{z_2}}; f_{22} = \frac{\overline{z_2} - z_2(x)}{\overline{z_2} - \underline{z_2}}; f_{21} + f_{22} = 1 \tag{3.12}$$

$$f_{31} = \frac{z_3(x) - \underline{z_3}}{\overline{z_3} - \underline{z_3}}; f_{32} = \frac{\overline{z_3} - z_3(x)}{\overline{z_3} - \underline{z_3}}; f_{31} + f_{32} = 1 \tag{3.13}$$

As noticed in Eq. (3.6), matrix A is expressed according to the premise z_1 only, so after developing the premise z_1 using formula (3.8), matrix A is expressed by the Eq. 3.14.

$$A = f_{11}A_1 + f_{12}A_2 \tag{3.14}$$

where: $A_1 = \begin{bmatrix} \underline{z_1} & 0 & 0 \\ 0 & -R_L/L & -1/L \\ 0 & 1/C_{DC} & -1/R_{DC}C_{DC} \end{bmatrix}$; $A_2 = \begin{bmatrix} \overline{z_1} & 0 & 0 \\ 0 & -R_L/L & -1/L \\ 0 & 1/C_{DC} & -1/R_{DC}C_{DC} \end{bmatrix}$

To express A in terms of polytope's vertex matrices defined by the partitions of the premises z_1, z_2, and z_3, Both sides of Eq. (3.14) are multiplied by the unit term $(f_{21} + f_{22})(f_{31} + f_{32})$. This leads to rewriting this equation as:

$$\begin{aligned} A = &(f_{11}f_{21}f_{31} + f_{11}f_{21}f_{32} + f_{11}f_{22}f_{31} + f_{11}f_{22}f_{32})A_1 \\ &+ (f_{12}f_{21}f_{31} + f_{12}f_{21}f_{32} + f_{12}f_{22}f_{31} + f_{12}f_{22}f_{32})A_2 \end{aligned} \tag{3.15}$$

3.4 PI Controller Development

Supposing:

$\vartheta_1 = f_{11}f_{21}f_{31}; \vartheta_2 = f_{11}f_{21}f_{32}; \vartheta_3 = f_{11}f_{22}f_{31}; \vartheta_4 = f_{11}f_{22}f_{32}; \vartheta_5 = f_{12}f_{21}f_{31};$
$\vartheta_6 = f_{12}f_{21}f_{32}; \vartheta_7 = f_{12}f_{22}f_{31}; \vartheta_8 = f_{12}f_{22}f_{32}$

Then:

$$A = (\vartheta_1 + \vartheta_2 + \vartheta_3 + \vartheta_4)A_1 + (\vartheta_5 + \vartheta_6 + \vartheta_7 + \vartheta_8)A_2 \quad (3.16)$$

Similarly, matrix B is expressed according to the partitions of the premises z_1, z_2, and z_3 as follows:

$$B = (\vartheta_1 + \vartheta_5)B_1 + (\vartheta_2 + \vartheta_6)B_2 + +(\vartheta_3 + \vartheta_7)B_3 + (\vartheta_4 + \vartheta_8)B_4 \quad (3.17)$$

where:

$$B_1 = \begin{bmatrix} \overline{z_2} \\ z_3 \\ 0 \end{bmatrix}; B_2 = \begin{bmatrix} \overline{z_2} \\ z_3 \\ 0 \end{bmatrix}; B_3 = \begin{bmatrix} \overline{z_2} \\ z_3 \\ 0 \end{bmatrix}; \text{ and } B_4 = \begin{bmatrix} \overline{z_2} \\ z_3 \\ 0 \end{bmatrix} \quad (3.18)$$

Finally, by substituting Eq. (3.16) and (3.17) into state Eq. (3.5), the complex state model of the PV system can be reformulated in terms of the linear subsystems M_i as follows:

$$\dot{X} = \sum_{i=1}^{8} \vartheta_i M_i \quad (3.19)$$

where:

$M_1 = A_1X + B_1u; M_2 = A_1X + B_2u; M_3 = A_1X + B_3u; M_4 = A_1X + B_4u;$
$M_5 = A_2X + B_1u; M_6 = A_2X + B_2u; M_7 = A_2X + B_3u; M_8 = A_2X + B_4u$

3.4 PI Controller Development

3.4.1 dP/dV Feedback Control Strategy

The dP/dV feedback control method is a widely used technique for Maximum Power Point Tracking (MPPT) in photovoltaic (PV) systems.

The derivative dP/dV denotes the slope of the P-V curve. Otherwise, it is evident that the MPP is at the top of the P-V curve when the slope of dP/dV is zero. In the proposed method, $dP/dV = 0$ serves as a reference for the PI controller. Figure 3.2 depicts the block design of the dP/dV MPPT control, which calculates the slope of dP/dV and compares it

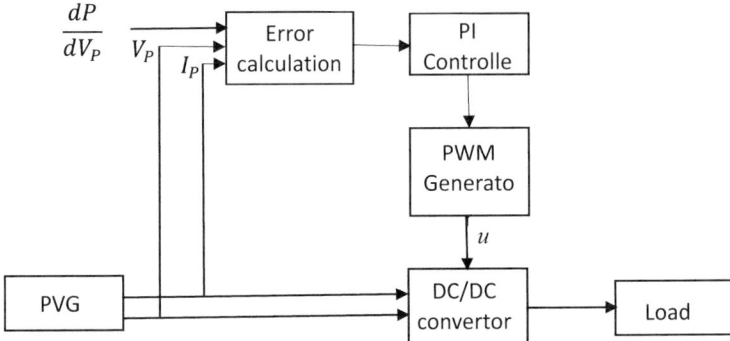

Fig. 3.2 Block design of the dP/dV MPPT control

to a fixed reference point of zero. The control method uses the produced error value to generate duty cycle sequences for the DC/DC converter switch gate [22].

Figure 3.2 shows that the controller input is the error value, representing the dP/dV slope, which is triggered by the PV system's feedback mechanism. As a result, the controller sends an output voltage as a control signal (duty cycle variation) to the DC/DC converter switch via the PWM generator.

The derivative dP/dV is expressed by the following equation:

$$\frac{dP}{dV} = I + \frac{dI}{dV} V \tag{3.20}$$

Thus,

$$\frac{dP}{dV} = 0 \Rightarrow \frac{dI}{dV} = -\frac{I_{mpp}}{V_{mpp}} = -\frac{1}{R_{st}} \tag{3.21}$$

where:

I_{mpp} is the current at MPP;
V_{mpp} is the voltage at MPP;
thereby, around the MPP, this derivative can be expressed by the following expression:

$$\frac{dP}{dV} = I - \frac{1}{R_{st}} V \tag{3.22}$$

3.4.2 Partial PI Controller Synthesis

Each of the eight models M_i predefined in Sect. 3.3, represents the behavior of the studied PV system within a specific area of its operating space. Consequently, developing a

3.4 PI Controller Development

performance controller for the partial model M_i facilitates performance control of the PV system within a localized operating zone. In this section, we focus on synthesizing eight partial PI controllers, specifically designed to ensure that the partial models track the set point $\frac{dP_i}{dV_{Pi}} = 0$.

With P_i and V_{pi} are the power and the voltage of the PV system when it works in the specific area corresponding to the partial model M_i.

By applying the formula (3.22) to the model M_i it results in the Eq. (3.23)

$$\frac{dP_i}{dV_{Pi}} = I_{Pi} - \frac{1}{R_{sti}} V_{Pi} \tag{3.23}$$

where: $\frac{1}{R_{sti}} = \frac{I_{mppi}}{V_{mppi}}$

with I_{mppi} and V_{mppi} are the current and the voltage related to M_i at the MPP, respectively.

The derivative $\frac{dP_i}{dV_{Pi}}$ appears as an addition to the current I_{Pi} and the voltage V_{pi} scaled by $\left(\left(-\frac{1}{R_{sti}}\right)\right)$. Thus, to cancel $\frac{dP_i}{dV_{Pi}}$ we must adjust I_{Pi} to I_{mppi} and $\left(Q = -\frac{1}{R_{sti}} V_{Pi}\right)$ to $\left(Q_m = -\frac{1}{R_{sti}} V_{mppi}\right)$.

This requires the design of a PI corrector to effectively control and stabilize simultaneously the current I_{Pi} and the quantity Q_i control loops. As demonstrated by [23], the PVG current, presented locally by I_{Pi}, can be regulated by managing the inductor current, presented locally by I_{Li}, while the control of Q_i is achieved through the regulation of V_{Pi}.

The M_i state model can be expressed in the following matrix form:

$$\begin{bmatrix} \dot{x}_{1i} \\ \dot{x}_{2i} \\ \dot{x}_{3i} \end{bmatrix} = \begin{bmatrix} \tilde{z}_1 & 0 & 0 \\ 0 & -R_L/L & -1/L \\ 0 & 1/C_{DC} & -1/R_{DC}C_{DC} \end{bmatrix} \begin{bmatrix} x_{1i} \\ x_{2i} \\ x_{3i} \end{bmatrix} + \begin{bmatrix} \tilde{z}_2 \\ \tilde{z}_3 \\ 0 \end{bmatrix} u_i \tag{3.24}$$

where:

- x_{1i}, x_{2i}, and x_{3i} are the state vector components related to the model M_i which represent the contributions of the partial model M_i to the PVG voltage, the inductor current, and the load voltage, respectively.
- $\tilde{z}_1$, $\tilde{z}_2$, and $\tilde{z}_3$ are the extremum (maximum or minimum) values of z_1, z_2, and z_3 respectively.

The vector $[x_{1i} x_{2i} x_{3i}]$ also denoted by $\begin{bmatrix} V_{pi} I_{Li} V_{DCi} \end{bmatrix}$ is expressed in Laplace space using the Eq. (3.24).

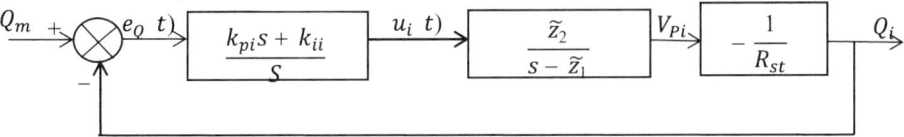

Fig. 3.3 Control loop of Q_i

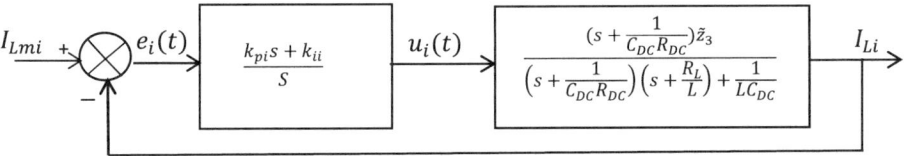

Fig. 3.4 Control loop of I_{Li}

$$\begin{bmatrix} x_{i1} \\ x_{i2} \\ x_{i3} \end{bmatrix} = \begin{bmatrix} V_{Pi} \\ I_{Li} \\ V_{DCi} \end{bmatrix} = \begin{bmatrix} \dfrac{\tilde{z}_2}{s-\tilde{z}_1} \\ \dfrac{\left(s+{}^1/C_{DC}R_{DC}\right)\tilde{z}_3}{\left(s+{}^1/C_{DC}R_{DC}\right)\left(s+{}^{R_L}/L\right)+{}^1/LC_{DC}} \\ \dfrac{\tilde{z}_3}{C_{DC}\left(s+{}^1/C_{DC}R_{DC}\right)\left(s+{}^{R_L}/L\right)+{}^1/L} \end{bmatrix} u_i \quad (3.25)$$

Building upon the last equation, the control loops for Q_i and I_{Li} are derived and illustrated in Figs. 3.3 and 3.4, respectively.

The closed-loop transfer functions $\frac{Q}{Q_m}$, and $\frac{I_{Li}}{I_{Lmi}}$ are given by Eqs. (3.26) and (3.27), respectively:

$$\frac{Q}{Q_m} = \frac{V_{Pi}}{V_{mppi}} = \frac{(k_{pi}s + k_{ii})\tilde{z}_2}{(k_{pi}s + k_{ii})\tilde{z}_2 - sR_{sti}(s - \tilde{z}_1)} \quad (3.26)$$

$$\frac{I_{Li}}{I_{Lmi}} = \frac{(k_{pi}s + k_{ii})\left(s + \frac{1}{C_{DC}R_{DC}}\right)\tilde{z}_3}{(k_{pi}s + k_{ii})\left(s + \frac{1}{C_{DC}R_{DC}}\right)\tilde{z}_3 + s\left(s + \frac{1}{C_{DC}R_{DC}}\right)\left(s + \frac{R_L}{L}\right) + \frac{s}{LC_{DC}}} \quad (3.27)$$

Our goal is to determine the values of regulator gains, k_{pi} and k_{ii} that ensure effective control and stability of both the I_{Li} and Q_i control loops. To achieve this, we calculate these gains by applying Routh criterion to the characteristic polynomial D_Q of the transfer function (3.26). Subsequently, we establish the complementary conditions that these gains must satisfy to guarantee the stability of $\frac{I_{Li}}{I_{Lmi}}$ the control loop.

The characteristic polynomial D_Q is given by:

$$D_Q = R_{sti}s^2 - (k_{pi}\tilde{z}_2 + R_{sti}\tilde{z}_1)s - k_{ii}\tilde{z}_2 \quad (3.28)$$

3.4 PI Controller Development

By applying Routh criterion to D_Q, the following conditions of stability are obtained:

- $C_1 : k_{pi} < -R_{sti}\frac{\tilde{z}_1}{\tilde{z}_2}$;
- $C_2 : k_{ii} > 0$.

Furthermore, by applying Routh criterion to the characteristic polynomial D_i derived from the transfer function $\frac{I_{Li}}{I_{Lmi}}$ leads to additional stability conditions $C_3, C_4, \text{ and } C_5$. The characteristic polynomial D_i is expressed as:

$$D_i = s^3 + \left(k_{pi}\tilde{z}_3 + \frac{R_L + L}{LC_{DC}R_{DC}}\right)s^2 + \left(k_{ii}\tilde{z}_3 + \frac{Lk_{pi}\tilde{z}_3 + R_L + R_{DC}}{LC_{DC}R_{DC}}\right)s + \frac{k_{ii}\tilde{z}_3}{R_{DC}C_{DC}} \tag{3.29}$$

The stability conditions are as follows:

- $C_3 : k_{pi} > -\frac{L+R_L}{\tilde{z}_3 L R_{CD} C_{DC}}$;
- $C_4 : k_{ii} > -\dfrac{\frac{(L+R_L)(\tilde{z}_3 L k_{pi}+R_L+R_{DC})}{(LR_{DC}C_{DC})^2} + \frac{\tilde{z}_3 k_{pi}(L\tilde{z}_3 k_{pi}+R_L+R_{DC})}{LR_{DC}C_{DC}} + \frac{\tilde{z}_3 k_{pi}}{C_{DC}R_{DC}}}{\tilde{z}_3^2 k_{pi} + \frac{L+R_L}{LR_{DC}C_{DC}}\tilde{z}_3}$;
- $C_5 : k_{ii}\frac{\tilde{z}_3}{R_{DC}C_{DC}} > 0$.

To confirm that the errors e_Q, and e_i in both loops will eventually converge to zero and maintain steady state monitoring of the set points, the final value theorem is employed as a validating method.

$$\lim_{t \to +\infty} e_Q(t) = \lim_{s \to 0} se_Q(s) = \frac{(s^3 R_{st} - s^2 R_{sti}\tilde{z}_1)Q_m}{R_{sti}s^2 - (k_{pi}\tilde{z}_2 + R_{st}\tilde{z}_1)s - k_{ii}\tilde{z}_2} = 0 \tag{3.30}$$

$$\lim_{t \to +\infty} e_i(t) = \lim_{s \to 0} se_i(s) \frac{s^2\left(\left(s + \frac{R_L}{L}\right)\left(s + \frac{1}{C_{DC}R_{DC}}\right) + \frac{1}{LC_{DC}}\right)I_{Lm}}{s^3 + \left(k_{pi}\tilde{z}_3 + \frac{R_L+L}{LC_{DC}R_{DC}}\right)s^2 + \left(k_{ii}\tilde{z}_3 + \frac{Lk_{pi}\tilde{z}_3+R_L+R_{DC}}{LC_{DC}R_{DC}}\right)s + \frac{k_{ii}\tilde{z}_3}{R_{DC}C_{DC}}} = 0 \tag{3.31}$$

3.4.3 Establishment of the PI Controller for the PV System

In engineering, the system's modeling is often considered as a preliminary step. When dealing with complex and highly non-linear processes, the development of a multi-model structure is a rather powerful and efficient method of overcoming the difficulties encountered in controller design. This is due to the fact that the complex system model is replaced by a set of simple models, each describes the process in a particular operating space. A multi-model control strategy can be developed within this framework by utilizing the predefined multi-model structure. These strategies have been effectively implemented in a variety of disciplines, such as engineering and industrial processes [24], as well as chemistry and biology [25, 26]. In this work, we present the development of a sophisticated controller for the photovoltaic (PV) system depicted in Fig. 3.1.

A weighting function is assigned to each partial controller in the proposed control strategy, which dynamically reflects its level of relevance at any given time t. The overall control signal $u(t)$ is then calculated by combining the weighted contributions of these partial controllers, as expressed by the following formula:

$$u(t) = K_P e(t) + K_I \int e(t) dt \qquad (3.32)$$

where:

$$K_P = \sum_{i=1}^{8} \vartheta_i k_{pi}, \; K_I = \sum_{i=1}^{8} \vartheta_i k_{ii}, \; and \; e(t) = -\frac{dP(t)}{dV_P(t)}.$$

The coefficients ϑ_i, previously set in Sect. 3.1, change over time. Thus, the global control applied to the PV system is termed adaptive control in the sense that the global controller parameters vary when climatic conditions change. Figure 3.5 shows the block diagram of the developed command strategy.

3.4.4 Stability of the PI Controller

The parameters of the PI controller are derived by using the eight partial PI controllers, each of which commands a linear model reflecting the PV system in an operating zone. The stability conditions for each linear model's PI controller are well established. However, this leads to a crucial question: How can we ensure that the global PI controller, which incorporates all partial controllers, can preserve overall system stability?

To address this concern and test the global PI controller's ability to maintain PV system stability, we can apply the simplified Borne-Gentina criterion. This criterion provides a clear and efficient method to evaluate stability. It acts as a guiding support to ensure that the global PI controller effectively maintains the stability of the controlled PV system. The simplified criterion of Borne-Gentina is stated as follows:

If there exists a matrix $A^* = \left\{ a_{i,j}^*(.) \right\}$ such that the non-constant elements are isolated in a single row or column and verifying:

$a_{i,i}^*(.) \geq a_{i,i}(.)$ and $a_{i,j}^*(.) \geq |a_{i,j}(.)|, \forall i \neq j$ where $a_{i,j}(.)$ are the elements of the matrix $A(.)$

Then, a sufficient condition for stability of the initial system is that the principal minors of the matrix A^* called the majorant matrix of $A(.)$ are of alternating signs, the first being negative, this means:

3.4 PI Controller Development

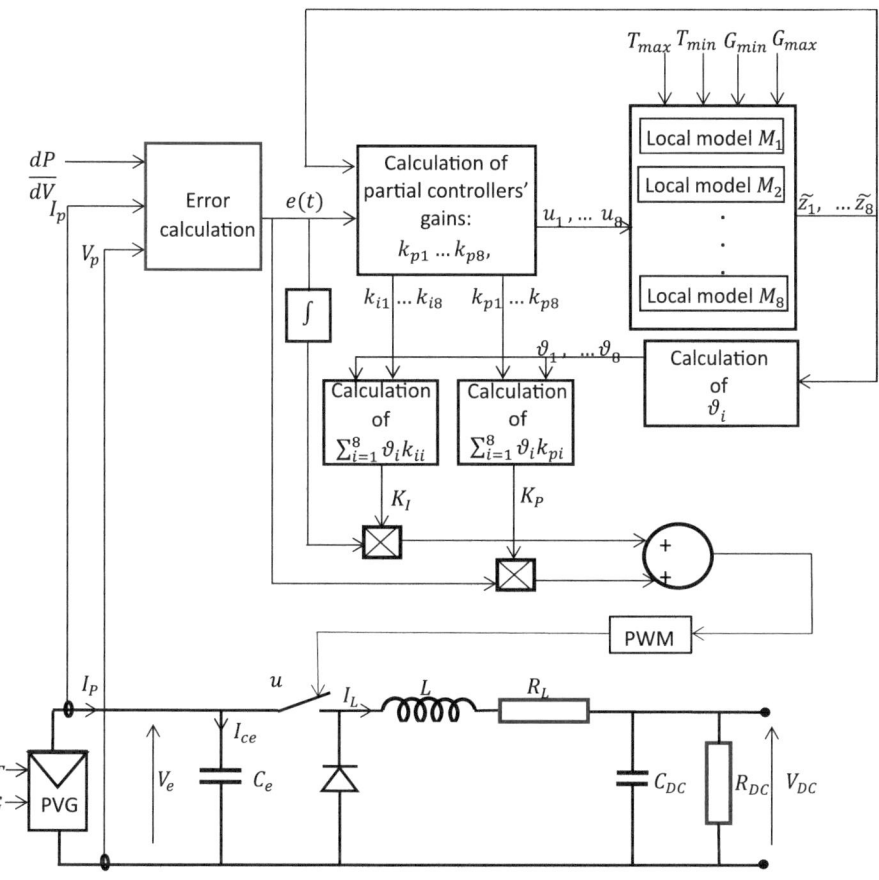

Fig. 3.5 Block diagram of the developed command strategy

$$a^*_{1,1}(.) < 0; \quad \begin{vmatrix} a^*_{1,1} & a^*_{1,2} \\ a^*_{2,1} & a^*_{2,2} \end{vmatrix} > 0; \quad (-1)^n \begin{vmatrix} a^*_{1,1} & \cdots & a^*_{1,n} \\ \vdots & \vdots & \vdots \\ a^*_{n,1} & \cdots & a^*_{n,n} \end{vmatrix} > 0$$

- **Stability of the voltage (V_P) control loop of the global system**

The control of the global system voltage is governed by the following closed-loop transfer function:

$$\frac{V_P}{V_{Pm}} = \frac{(K_P s + K_I) z_2}{(K_P s + K_I) z_2 - s R_{st}(s - z_1)} \qquad (3.33)$$

From this transfer function, we deduce the following state equation:

$$\dot{x} = Dx + Ev \quad (3.34)$$

with:

$$v = V_{Pm}; D = \begin{bmatrix} 0 & 1 \\ -\alpha_0 & -\alpha_1 \end{bmatrix}; E = \begin{bmatrix} 0 \\ 1 \end{bmatrix}; \alpha_0 = -\frac{K_I}{R_{st}};$$

$$\alpha_1 = -\frac{R_{st}z_1 + K_P z_2}{R_{st}}; z_1(x) = \frac{I_P}{C_e V_P}, z_2(x) = -\frac{I_L}{C_e}.$$

we suppose the transition matrix $P = \begin{bmatrix} 1 & 0 \\ p_1 & 1 \end{bmatrix}$ with p_1 is the pole of the model M_i, which verifies the condition: $|p_1| = \max(|p_i|) i = 1 \ldots 8$

$$P^{-1}AP = \begin{bmatrix} p_1 & 1 \\ -(\alpha_0 + \alpha_1 p_1 + p_1^2) & -(\alpha_1 + p_1) \end{bmatrix} = \begin{bmatrix} p_1 & 1 \\ \gamma_1 & \gamma_2 \end{bmatrix} = A_F$$

The following matrix M_F is a majorant matrix of A_F (consequently is a majorant of A).

$$M_F = \begin{bmatrix} p_1 & 1 \\ \max(\gamma_1) & \max(\gamma_2) \end{bmatrix}; \text{ with } \max(\gamma_1) = |\overline{\alpha}_0 + \overline{\alpha}_1 p_1 + p_1^2|; \max(\gamma_2) = -\overline{\alpha}_1 - p_1.$$

$\overline{\alpha}_0$ and $\overline{\alpha}_1$ are the maximum values of α_0 and α_1

$$max(\gamma_1) = K_{Imax} * \frac{\overline{z_2}}{R_{st}} + \frac{R_{st}\overline{z_1} + K_{Pmax}\overline{z_2}}{R_{st}} p_1 - p_1^2; max(\gamma_2) = \frac{R_{st}\overline{z_1} + K_{Pmax}\overline{z_2}}{R_{st}} - p_1$$

where $K_{Imax} = \max(k_{ii}); K_{Pmax} = \max(k_{pi}) i = 1 \ldots 8$.

One sufficient condition for stability is:

$p_1 < 0$; and $\begin{vmatrix} p_1 & 1 \\ \max(\gamma_1) & \max(\gamma_2) \end{vmatrix} > 0.$

Since p_1 is a pole of the stable local model M1, then $p_1 < 0$ is satisfied.
On the other hand:

$$\begin{vmatrix} p_1 & 1 \\ \max(\gamma_1) & \max(\gamma_2) \end{vmatrix} = -K_{Imax} * \frac{\overline{z_2}}{R_{st}},$$

where K_{Imax} is constrained to be positive by the predefined condition C_2, and $\overline{z_2}$ is negative consequently: $-K_{Imax} * \frac{\overline{z_2}}{R_{st}} > 0$.

From this, we deduce that the established PI controller is stabilizing for the V_P voltage control loop of the global system.

3.4 PI Controller Development

- **Stability of the current (I_L) control loop of the global system:**

The stability of the control for the global system current I_L is similarly examined using the same method. The closed-loop transfer function corresponding to the global system current I_L is expressed as:

$$\frac{I_L}{I_{Lm}} = \frac{(K_{PS} + K_I)\left(s + \frac{1}{C_{DC}R_{DC}}\right)z_3}{(K_{PS} + K_I)\left(s + \frac{1}{C_{DC}R_{DC}}\right)z_3 + s\left(s + \frac{1}{C_{DC}R_{DC}}\right)\left(s + \frac{R_L}{L}\right) + \frac{s}{LC_{DC}}} \tag{3.35}$$

We deduce the following state equation:

$$\dot{x} = D_1 x + E_1 \vartheta_1$$

$$D_1 = \begin{bmatrix} 0 & 1 & 0 \\ 0 & 0 & 1 \\ -\beta_0 & -\beta_1 & -\beta_2 \end{bmatrix}; E_1 = \begin{bmatrix} 0 \\ 0 \\ 1 \end{bmatrix}; \vartheta_1 = I_{Lm}; \beta_0 = \frac{z_3 K_I}{C_{DC}R_{DC}};$$

$$\beta_1 = \frac{z_3 K_P}{C_{DC}R_{DC}} + z_3 K_I + \frac{R_{DC} + R_L}{LC_{DC}R_{DC}}; \beta_2 = z_3 K_P + \frac{R_L}{L} + \frac{1}{C_{DC}R_{DC}};$$

We consider the transition matrix $Q = \begin{bmatrix} 1 & 1 & 0 \\ P_1 & P_2 & 0 \\ P_1^2 & P_2^2 & 1 \end{bmatrix}$,

Where: $|p_1| = \min(|p_i|) i = 1 \ldots 8$ and $|p_1| < |p_2| < \min(|p_i|); j \neq i$

$$Q^{-1}A_1Q = \begin{bmatrix} p_1 & 0 & -\frac{1}{p_1 - p_2} \\ 0 & p_2 & \frac{1}{p_1 - p_2} \\ -(\beta_0 + \beta_1 p_1 + \beta_2 p_1^2 + p_1^3) & -(\beta_0 + \beta_1 p_2 + \beta_2 p_2^2 + p_2^3) & -(\beta_2 + p_1 + p_2) \end{bmatrix} = Q_F$$

The following matrix Q_M is a majorant matrix of Q_F:

$$Q_M = \begin{bmatrix} p_1 & 0 & \left|-\frac{1}{p_1-p_2}\right| \\ 0 & p_2 & \left|\frac{1}{p_1-p_2}\right| \\ \max(\delta_0) & \max(\delta_1) & \max(\delta_2) \end{bmatrix}$$

$$\max(\delta_0) = \frac{\overline{z_3}K_{Imax}}{C_{DC}R_{DC}} + \left(\frac{\overline{z_3}K_{Pmax}}{C_{DC}R_{DC}} + \overline{z_3}K_{Imax} + \frac{R_{DC}+R_L}{LC_{DC}R_{DC}}\right)p_1 + (\overline{z_3}K_{Pmax} + \frac{R_L}{L} + \frac{1}{C_{DC}R_{DC}})p_1^2 + p_1^3$$

$$\max(\delta_1) = \frac{\overline{z_3}K_{Imax}}{C_{DC}R_{DC}} + \left(\frac{\overline{z_3}K_{P\max}}{C_{DC}R_{DC}} + \overline{z_3}K_{Imax} + \frac{R_{DC}+R_L}{LC_{DC}R_{DC}}\right)p_2 + (\overline{z_3}K_{P\max} + \frac{R_L}{L}$$

$$+ \frac{1}{C_{DC}R_{DC}})p_2^2 + p_2^3$$

$$\max(\delta_2) = -\left(\overline{z_3}K_{Pmax} + \frac{R_L}{L} + \frac{1}{C_{DC}R_{DC}} + p_1 + p_2\right)$$

A sufficient condition of stability is:

$$p_1 < 0; \begin{vmatrix} p_1 & 0 \\ 0 & p_2 \end{vmatrix} > 0; \begin{vmatrix} p_1 & 0 & \left|-\frac{1}{p_1-p_2}\right| \\ 0 & p_2 & \left|\frac{1}{p_1-p_2}\right| \\ \max(\delta_0) & \max(\delta_1) & \max(\delta_2) \end{vmatrix} < 0$$

This condition is verified because

$$p_1 < 0; \begin{vmatrix} p_1 & 0 \\ 0 & p_2 \end{vmatrix} = p_1p_2 > 0; \begin{vmatrix} p_1 & 0 & \left|-\frac{1}{p_1-p_2}\right| \\ 0 & p_2 & \left|\frac{1}{p_1-p_2}\right| \\ \max(\delta_0) & \max(\delta_1) & \max(\delta_2) \end{vmatrix} = -\frac{z_3K_I}{C_{DC}R_{DC}} < 0$$

We deduce that the established PI controller is stabilizing for the I_L current control loop of the global system.

3.5 Results and Discussions

For simulation and analysis in this work, we used the PV module SPM (P) 250W SPM (P) 250W, which has the parameters given in Table 3.2.

Table 3.2 Parameters of the PV module SPM (P) 250W for (STC) and (NOCT) conditions

	Standard testing conditions (STC):	Normal operating cell temperature conditions (NOCT)
Peak power Pmax	250W	181.3W
Maximum power voltage Vmp	30.2 V	27.93 V
Maximum power current Imp	8.28A	6.49A
Open circuit voltage Voc	36.6 V	34.55 V
Short circuit current Isc	8.83A	7.03A
Temperature coefficient of Voc $\beta_0 = -0.32\%$		
Temperature coefficient of Isc $\alpha_0 = 0.052\%$		
Ns = 60		
NOCT = $45 \mp 2\,°C$.		

3.5 Results and Discussions

Because environmental factors naturally affect photovoltaic (PV) module performance, their parameters change significantly with temperature and irradiance levels. The complex relationships between module parameters and environmental factors are modeled using mathematical expressions to precisely capture these dependencies and predict PV module behavior in practical situations. Researchers and engineers can comprehend, model, and optimize PV module performance under a variety of climatic conditions, thanks to these expressions, which serve as the basis for PV module modeling. By including these equations, the modeling procedure guarantees a thorough depiction of how environmental factors, such as temperature swings and fluctuating sunlight intensity, affect parameters such as current, voltage, and power output. The equations are given as follows [27, 28]:

$$V_P = V_{mp}\left[1 + 0.0539 \ln\left(\frac{G}{G_0}\right)\right] + \beta_0(T_C - T_{C0}) \quad (3.36)$$

$$I_{ph} = \frac{G}{G_0}(I_{SC} + \alpha_0(T_C - T_{C0})) \quad (3.37)$$

$$I_S = I_{S0}\left(\frac{T_C}{T_{C0}}\right)^3 e^{\frac{qE_g}{nKN_S}\left(\frac{1}{T_{C0}} - \frac{1}{T_C}\right)} \quad (3.38)$$

$$I_{S0} = \frac{I_{SC}}{e^{\frac{qV_{OC}}{nKN_S T_{C0}}} - 1} \quad (3.39)$$

$$T_c = T_a + \left[\frac{NOCT - 20}{800}\right]G \quad (3.40)$$

$$E_g = E_{g0}(1 - 0.00026(T_C - T_{C0})) \quad (3.41)$$

$$R_P = \frac{G_0}{G}R_{P0} \quad (3.42)$$

$$R_S = R_{S0} \quad (3.43)$$

where: T_a: the ambient temperature; T_{C0}: Temperature of the PV module at STC; n: The ideality factor which depends on PV cell technology and can be chosen in [29], E_g is the bang-gap energy of the semiconductor used in the cell, and E_{g0} is the electron bandgap and under STC conditions.

In this work, we chose to perform simulations in Matlab environment by considering a sampling step of 0.0005 s, a range of 200–1000 W/m² for irradiance and a range of 0°C to 70°C for temperature. Characteristics I-V and P-V of the module, at different climatic conditions, are presented below in Figs. 3.6 and 3.7.

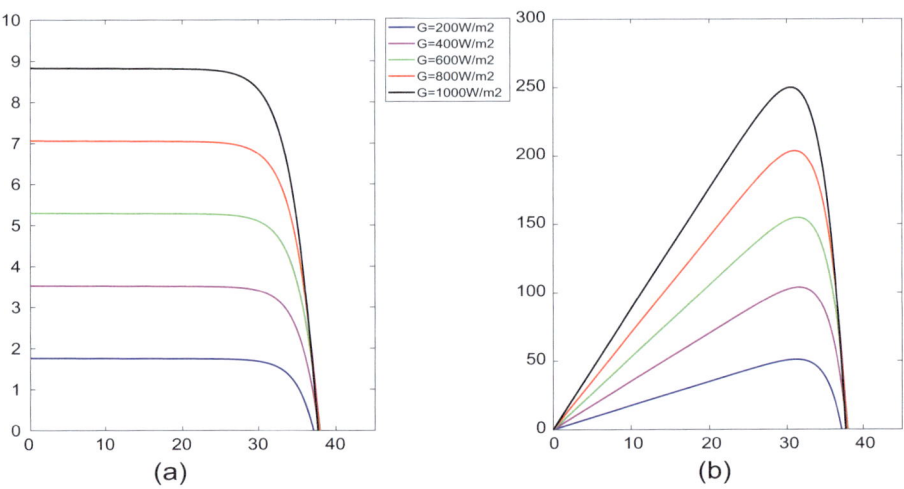

Fig. 3.6 Current-voltage characteristics (Fig. 3.6a) and power-voltage characteristics (Fig. 3.6b) under various irradiation and fixed temperature T = 25 °C

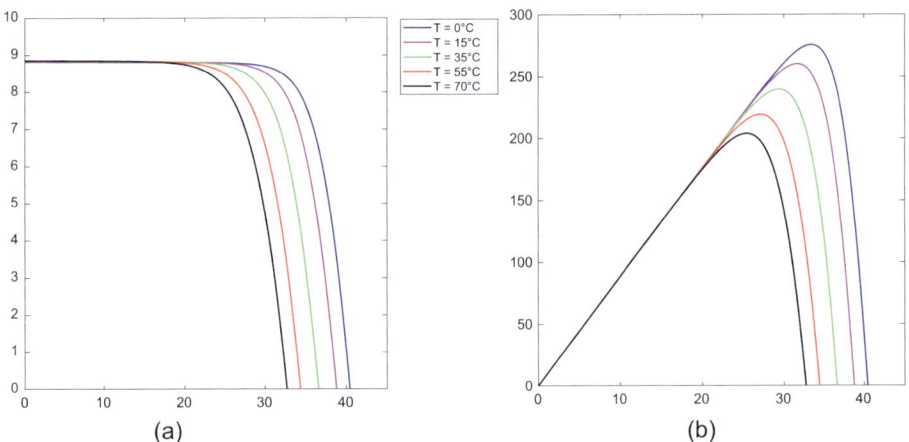

Fig. 3.7 Current-voltage characteristics (Fig. 3.7a) and power-voltage characteristics (Fig. 3.7b) under various temperature and fixed irradiation G = 1000 W/m²

3.5.1 Model Simulation

To assess the accuracy of the developed model, the error between the PV system voltage V_P and that produced by this model V_{mP} is simulated over a wide range of temperatures varying from 0 to 70 °C and irradiances varying 200–1000 W/m². The convertor

3.5 Results and Discussions

Table 3.3 Parameters of the buck convertor

Buck convertor parameters	Value
Input capacitor Ce	$47.10^{-3} F$
Inductor L	0.01H
Output capacitor C_{DC}	$47.10^{-3} F$
Resistor R_{DC}	$100 \, \Omega$
Resistor R_L	$1 \, \Omega$

parameters used in this simulation are determined using the formulas provided in [30, 31]. Table 3.3 presents these parameters, and Fig. 3.8 shows the simulation results.

We observe that the evolution of the error under a wide range of climatic conditions of irradiance and temperature decreases below 8.10^{-15}. These results highlight the model's excellent ability to generate the exact voltage V_p of the PV system. To further confirm the high quality of the developed model, it is important to assess its capacity to replicate the current I_L and the voltage V_{DC}. We proceed by simulating the error between the system's current I_L and the corresponding current I_{mL} produced by the model, as well as the error between V_{DC} and the corresponding voltage V_{mDC} produced by the model across a wide range of climatic conditions. Figures 3.9 and 3.10 represent the simulation results.

Results presented in Figs. 3.9 and 3.10 demonstrate that the evolution of errors under various climatic conditions of irradiance and temperature decreases below 8.10^{-16} and 4.10^{-18} respectively. These results emphasize the excellent capability of the model to

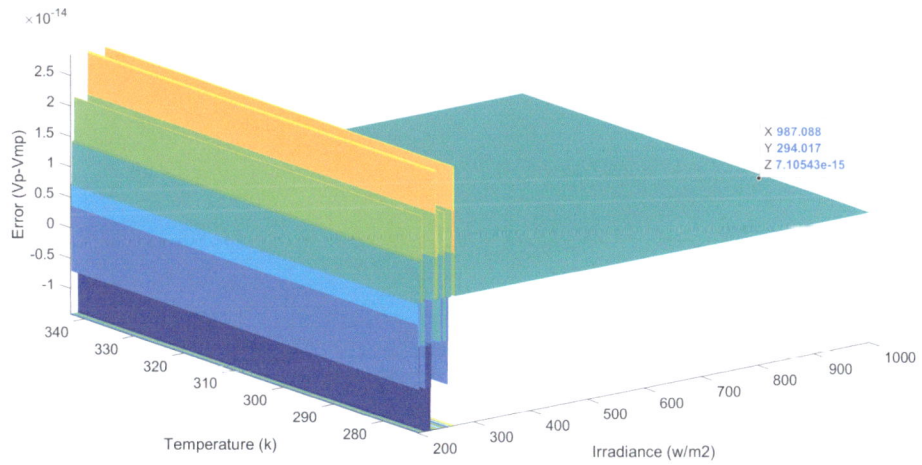

Fig. 3.8 Variation of the error between the PVG voltage and the model voltage across a range of irradiance from 200 to 1000 W/m^2 and temperature from 0 to 70 °C

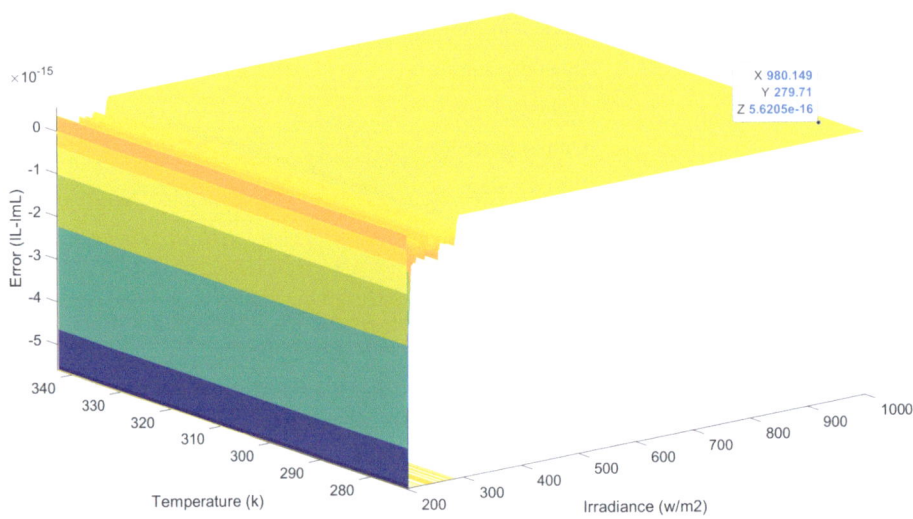

Fig. 3.9 Evolution of the error between the current I_L and the model's corresponding current I_{mL} across a range of temperatures from 0 to 70 °C and irradiances from 200 to 1000 W/m²

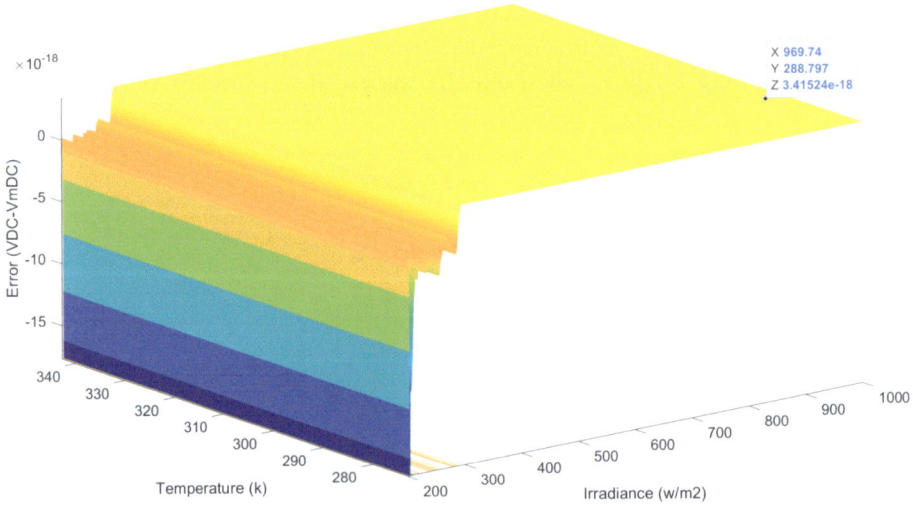

Fig. 3.10 Evolution of the error between the voltage V_{DC} and the model's corresponding voltage V_{mDC} across a range of temperatures from 0 to 70 °C and irradiances from 200 to 1000 W/m²

3.5 Results and Discussions

accurately replicate the behavior of the PV system, proving the high quality of the developed model. This analysis provides a definitive validation of the model's performance and accuracy under a wide range of operating conditions. With the outstanding quality of the developed model confirmed, the next step involves the simulation of the PI controller.

3.5.2 Controller Simulation

To assess the effectiveness of the developed controller, the first step involves simulating the voltage V_P, the current I_P, and the power P outputs generated by the controlled PV system under standard conditions of temperature and irradiance, and comparing the simulation findings with the maximum values of V_P, I_P, and P given on the datasheet. Figure 3.11 represents the evolution of the controlled PV system's voltage, current, and power under standard conditions of temperature and irradiance, thereby the evolution of the control signal u.

According to the results given in the above figure, it can be observed that in the stationary regime, the PVG voltage stabilizes at 30.19V, the current reaches 8.277A, and the power reaches 249.918W. The controller has succeeded in forcing the PV system to operate at the maximum power point without causing any ripple. This confirms the exceptional performance of the developed controller. Next, the controller's ability to track the maximum power point (MPP) will be evaluated under sudden changes in climatic

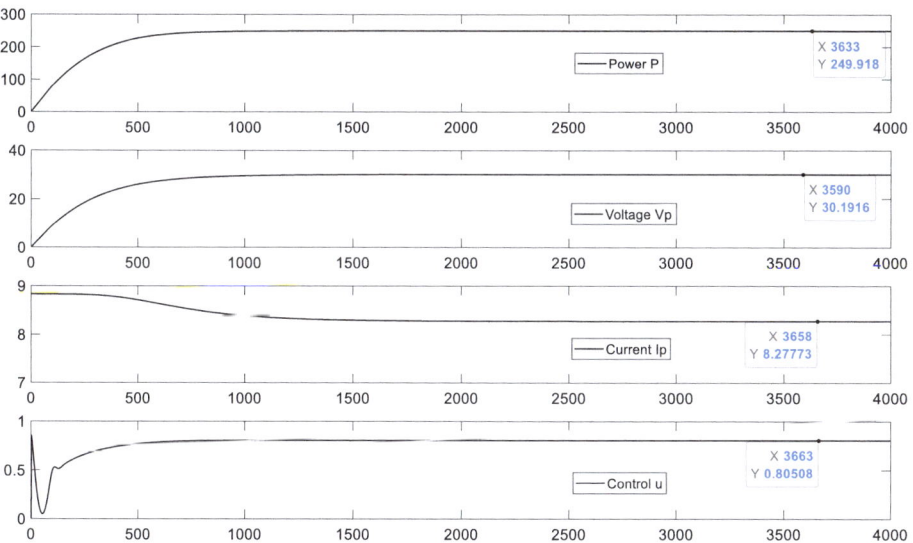

Fig. 3.11 Evolution of the PV system voltage, power, current, and control signal under standard conditions of temperature and irradiance

conditions. NOCT conditions have been selected for this test, as the maximum electrical power P is provided in the datasheet. This allows for a direct comparison between the datasheet values and the results achieved by the controlled system. The evolution of the PV system power, the variation of the controller gains and the evolution of the command signal (duty cycle) under the climatic condition change from (G = 1000 W/m^2, T = 25 °C) to (G = 800 W/m^2, T = 20 °C) are presented in Fig. 3.12.

The results in Fig. 3.12 demonstrate the ability of the suggested controller to precisely adjust the power output in response to environmental changes without creating oscillations. Additionally, the controller stabilizes the power signal in a short duration t = 0.5 s, demonstrating a ripple-free performance. The lack of ripple is confirmed in both the control and gain signals. As a result, these findings validate the controller's ability to deliver excellent tracking performance. To critically evaluate the controller's

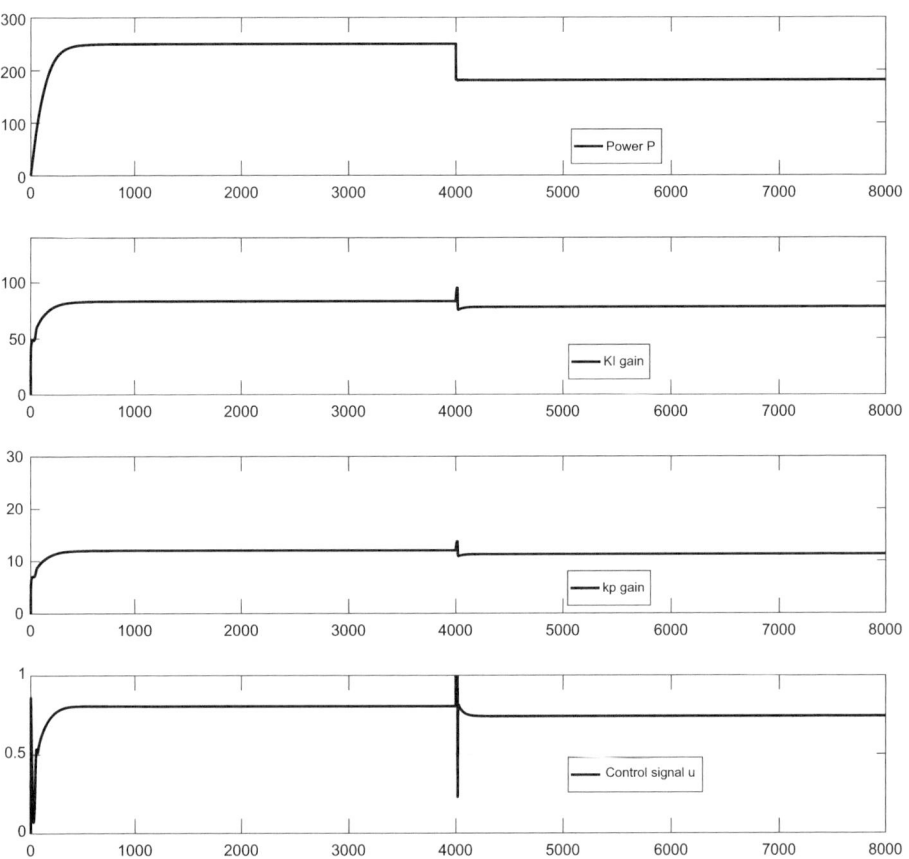

Fig. 3.12 Evolutions of the PV system power, the controller gains, and the control signal under climatic conditions change from (T = 25 °C; G = 1000 W/m^2) to (T = 20 °C; G = 800 W/m^2)

3.5 Results and Discussions

effectiveness, a comparative analysis with a conventional MPPT controller was conducted. For this objective, an MPPT controller based on the Perturb and Observe (P&O) method was implemented under changing climatic conditions from (T = 25 °C; G = 1000 W/m^2) to (T = 20 °C; G = 800 W/m^2). The power outputs of the photovoltaic system, managed by both the P&O method and the elaborated PI controller, are shown in Fig. 3.13. To provide a clearer comparison, sections A and B of the curves have been magnified and are highlighted within the same figure.

Results given in Fig. 3.13, allow to conclude on the performances of the two controllers. These results highlight significant differences in their performance. The proposed PI controller demonstrates superior stability, showing no overshoot or undershoot during transients; however the P&O-based controller demonstrates an overshoot of 25.55% an undershoot of 12.25%. Additionally, the proposed PI controller cancels definitely the ripple in the power signal, contrary to the P&O controller, which displays a ripple of $\Delta P_{max} = 0.9$ w. The PI controller effectively tracks the new maximum power point (MPP) without causing oscillatory behavior during transients, although the P&O controller shows significant oscillations before reaching the new MPP. Regarding settling time, the proposed PI controller reaches and remains within 2% of its final value at t = 0.5s, much faster than the P&O controller, which settles at t = 0.9s. Additionally, the steady-state error is significantly lower for the PI controller, achieving 0.15w and 0.1w under two

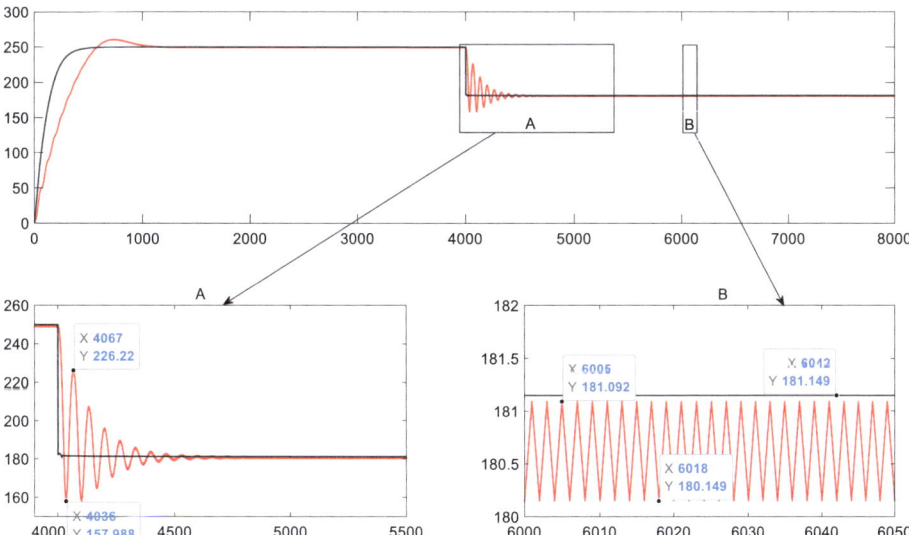

Fig. 3.13 Power variation of the PV system controlled by the MPPT controller based on P&O method (Pink curve) and that of the same system controlled by the proposed PI controller (Blue curve) in the case of climatic conditions change from (T = 25 °C; G = 1000 W/m^2) to (T = 20 °C; G = 800 W/m^2), and too zoomed parts A and B of these curves

climatic conditions, compared to 1w and 1.3w for the P&O controller under the same different conditions. This evaluation demonstrates the proposed PI controller's capacity to give remarkable tracking performance while also establishing that it is considerably more reliable than the P&O-based MPPT controller.

3.6 Conclusion

This chapter elaborates on a performance and adjustable proportional integral (PI) controller aimed at maximum power point tracking for photovoltaic PV systems. The implementation of an adaptive multi-model controller is performed based on a multi-model structure that aims to replicate the behavior of the PV system under varying climatic conditions of temperature and irradiance. The multi-model structure was obtained by employing a polytopic transformation of the PV system's state model, which allows the complex state model to be reorganized as the sum of eight linear subsystems, each of which is assigned by a nonlinear weighting function. This transformation facilitates obtaining an enhanced representation of the dynamics of the PV system. In order to validate the developed multi-model structure, simulations were performed within a commercial SPM (P) 250W PV module working with a buck converter. These experiments confirmed that the error margin between real responses of the PV system and the output responses of the model was considerably small: less than 10^{-14}. This proves the developed model's capability of accurately describing the behavior of PV systems under different levels of irradiance and temperatures. The proposed PI controller was constructed to control the PV system using the feedback control strategy of dP/dV. Its proportional and integral gains were computed by merging the gains from partial PI controllers designed for each linear subsystem after weighting them by the nonlinear weighting functions obtained in the modeling stage. These controllers were developed to guarantee simultaneously the stability of the inductor current control loop and a specified model voltage function control loop.

In order to verify the performance of the proposed PI controller, a series of simulations was conducted. These included the most common comparisons with the P&O based MPPT controller using different environmental test profiles that included changes in irradiance and temperature. The controller performance was evaluated in terms of achieving the maximum power point, maintaining stabilization at the MPP, meeting the settling time, and the transients in the system from overshoot, undershoot, ripple, and oscillation during transients. The results clearly indicate the superior performance of the proposed PI controller. It accurately converges to MPP within a settling time of 0.5s, and it does all this in a non-oscillatory manner during transients. Also, the energy tracking (no ripple) in an output isolates a perfect offsetting of the power variations from PV. In general, though, the P&O MPPT controller suffered more from its weaknesses, generation of ripples, significant transients' oscillations, and experienced an overshoot of 25.55% before stabilizing.

The findings revealed that not only is the suggested PI controller accurate, stable, and efficient, but it is also easy to install, indicating that it outperforms the P&O-based MPPT controller by far. With the micro-computing platform compatible technique and low cost, the proposed PI controller is also envisaged as a viable option for marketable PV systems, providing better reliability and performance under real working conditions.

References

1. Manisha, Penkey, M. Kumari, R.K. Sahdev, S. Tiwari, A review on solar photovoltaic system efficiency improving technologies. Appl. Sol. Energy **58**, 54–75 (2022)
2. A.C. Lazaroiu, M. Gmal Osman, C.V. Strejoiu, G. Lazaroiu, A comprehensive overview of photovoltaic technologies and their efficiency for climate neutrality. Sustainability **15**(23) (2023)
3. Aiko Solar. Neostar Series Panels: Advanced N-Type ABC Technology. AIKO Unveils Next-Generation High-Efficiency N-type ABC Solar Modules—AIKO, Find Your Power (2024)
4. N. Mensia, M. Talbi, M. Bouaicha, New adaptive PI controller for photovoltaic systems. Iran. J. Sci. Technol. Trans. Electr. Eng. **48**, 1099–1110 (2024)
5. M. Talbi1, N. Mensia, H. Ezzaouia, New photovoltaic module model and a comparative study of MPPT control techniques based on neural networks. Light. Eng. **29**(1), 69–76 (2021)
6. M. Talbi, N. Mensia, H. Ezzaouia, Modeling of a PV panel and application of maximum power point tracking command based on ANN. Int. Arab. J. Inf. Technol. **18**(4) (2021)
7. C. Pardhi, K. Khare, A. Choubey, Impact of irradiance and temperature on electrical parameters of polycrystalline photovoltaic module. Int. J. Recent. Technol. Eng. (IJRTE) **13**(2), 12–20 (2024)
8. D. Toumi, D. Benattous, A. Ibrahim, B. Tarek, Maximum power point tracking of photovoltaic array using fuzzy logic control. Int. J. Power Electron. Drive Syst. (IJPEDS) **13**(4), 2440 (2022)
9. S. Saravanan, N.R. Babu, Maximum power point tracking algorithms for photovoltaic system—a review. Renew. Sustain. Energy Rev. **57**, 192–204 (2016)
10. N. Karami, N. Moubayed, R. Qutbib, General review and classification of different MPPT techniques. Renew. Sustain. Energy Rev. **68**, 1–18 (2017)
11. V.C. Tella, B. Agili, M. He, Advanced MPPT control algorithms: a comparative analysis of conventional and intelligent techniques with challenges. Eur. J. Electr. Eng. Comput. Sci. **8**(4), 6–20 (2024)
12. B. Bora, S.K. Gudey, Performance and analysis of adaptable PI and SMC for a hybrid PV-battery system, in *IEEE International WIE Conference on Electrical and Computer Engineering (WIECON-ECE)*, pp. 26–27 (2020)
13. D. Abbes, C.H. Gerard, M. Andre, R. Benoit Modeling and simulation of a photovoltaic system: an advanced synthetic study, in *3rd International Conference on Systems and Control*, pp. 29–31 (2013)
14. K. Dhruv, K.P.S. Kumar, V. Rana, A nonlinear PID controller based novel maximum power point tracker for PV systems. J. Franklin Inst. **355**(16), 7827–7864 (2018)
15. B.Y. Zhao, Z.G. Zhao, Y. Li, R.Z. Wang, R.A. Taylor, An adaptive PID control method to improve the power tracking performance of solar photovoltaic air-conditioning systems. Renew. Sustain. Energy Rev. **113**, 109250 (2019)
16. D. Klemen, B. Peter, S. Klemen, S. Sebastijan, Proportional integral controllers performance of a grid-connected solar PV system with particle swarm optimization and Ziegler-Nichols tuning method. Energies **14**, 2516 (2021)

17. K. Miyazaki, K.W. Bowman, K. Yumimoto, T. Walker, K. Sudo, Evaluation of a multi-model, multi-constituent assimilation framework for tropospheric chemical reanalysis. Atmos. Chem. Phys. **20**(2), 931–967 (2020)
18. L.M. Juana, A.A.S. Edgar, O.N.M. Manuel, A.T. Javier, M.B. Enrique, Modeling growth on cannonball jellyfish Stomolophus meleagris based on a multi-model inference approach. Hydrobiologia **847**, 1399–1422 (2020)
19. K. Damian, S. Bozena, H. Anna, A heuristic and simulation hybrid approach for mixed and multi model assembly line balancing, in *Conference proceedings in Intelligent Systems in Production Engineering and Maintenance – ISPEM*, pp. 99–108 (2017)
20. S. Umashankar, K.P. Aparna, R. Priya, S. Suryanarayanan, Modeling and simulation of a PV system using DC-DC converter. Int. J. Latest Res. Eng. Technol. (IJLRET) **1**(2), 9–16 (2015)
21. M. Nawel, T. Mourad, B. Mongi, Modelling of photovoltaic water pumping system based on artificial intelligence. Adv. Model. Anal. B. **62**(1), 11–17 (2019)
22. M. Tadj, L. Chaib, A. Choucha, A.M. Aldaoudeyeh, A. Fathy, H. Rezk, M. Louzazni, A. El-Fergany, Enhanced MPPT-based fractional-order PID for PV systems using aquila optimizer. Math. Comput. Appl. **28**(5) (2023)
23. P. Han-Eol, S. Joong-Ho, A dP/dV feedback-controlled MPPT method for photovoltaic power system using II-SEPIC. J. Power Electron. **9**(4), 604–611 (2009)
24. M. Schwenzer, M. Ay, T. Bergs, D. Abel, Review on model predictive control: an engineering perspective. Int. J. Adv. Manuf. Technol. **117**, 1327–1349 (2021)
25. R. Argelaguet, D. Arnol, D. Bredikhin, Y. Deloro, B. Velten, J.C. Marioni, O. Stegle, MOFA+: a statistical framework for comprehensive integration of multi-modal single-cell data. Genome Biol. **21**(1), 2–17 (2020)
26. S.J. Qin, T.A. Badgwell, A survey of industrial model predictive control technology. Control. Eng. Pract. **11**(7), 733–764 (2003)
27. T. Mourad, N. Mensia, F. Krout, R. Chtourou, Matlab/Simulink and experimental studies of shading effect on a photovoltaic array. Int. J. Eng. Res. Technol. (IJERT) **6**(3), 588–593 (2017)
28. B. Habbati, Y. Ramdani, F. Moulay, A detailed modeling of photovoltaic module using nMATLAB. NRIAG J. Astron. Geophys. **3**, 53–6 (2014)
29. H.L. Tsai, C.S. Tu, Y.J. Su, Development of generalized photovoltaic model using MATLAB/Simulink, in *Proceedings of the World Congress on Engineering and Computer Science* (2008)
30. E. Koutroulis, K. Kalaitzakis, N. Voulgaris, Development of a microcontroller-based photovoltaic maximum point tracking control system. IEEE Trans. Power Electron. **16**(1), 46–54 (2001)
31. J.M. Enrique, E. Duran, M. Sidrach-de-Cardona, J.M. Andujar, Theoretical assessment of the maximum power point tracking efficiency of photovoltaic facilities with different converter topologies. Sol. Energy **81**, 31–38 (2007)

MIX
Papier aus verantwortungsvollen Quellen
Paper from responsible sources
FSC® C105338

If you have any concerns about our products,
you can contact us on
ProductSafety@springernature.com

In case Publisher is established outside the EU,
the EU authorized representative is:
**Springer Nature Customer Service Center GmbH
Europaplatz 3, 69115 Heidelberg, Germany**

Printed by Libri Plureos GmbH
in Hamburg, Germany